Good Questions

Great Ways to Differentiate Mathematics Instruction

Good Questions

Great Ways to Differentiate
Mathematics Instruction

MARIAN SMALL

Teachers College, Columbia University
New York and London

NATIONAL COUNCIL OF
TEACHERS OF MATHEMATICS

1906 Association Drive, Reston, VA 20191
www.nctm.org

www.nelson.com

Published simultaneously by Teachers College Press, 1234 Amsterdam Avenue, New York, NY 10027, and National Council of Teachers of Mathematics, 1906 Association Drive, Reston, VA 20191; distributed in Canada by Nelson Education, 1120 Birchmount Road, Toronto, ON, Canada M1K 5G4.

Credits: Teddy bear, boat, and frame (page 39) from *Nelson Math Focus 2* by Marian Small, page 74. Copyright © 2008. Reprinted with permission of Nelson Education Limited.

Text Design: Lynne Frost

Library of Congress Cataloging-in-Publication Data

Small, Marian.
    Good questions : great ways to differentiate mathematics instruction / Marian Small.
        p.   cm.
    Includes bibliographical references and index.
    ISBN 978-0-8077-4978-4 (pbk. : alk. paper)
        1. Mathematics—Study and teaching (Elementary).   2. Individualized instruction.
    3. Effective teaching.   I. Title.

    QA20.I53S63   2009
    372.7—dc22   2008054815

ISBN 978-0-8077-4978-4 (paper)
NCTM Stock Number: 13513

Printed on acid-free paper
Manufactured in the United States of America

16 15 14 13 12 11 10 09   8 7 6 5 4

Contents

Preface

IN 2001, I embarked on a long-term research study to investigate how mathematical understanding develops in students through the elementary years. The project was housed at the University of New Brunswick and the University of Toronto, and it was supported by Nelson Education Ltd. The work ultimately led to a professional development program called PRIME, which explores student development in mathematics. The research involved testing over 12,000 kindergarten through grade 7 students across Canada.

The research resulted, in part, in the creation of developmental maps that show the stages students move through in each of the strands of mathematics articulated by the National Council of Teachers of Mathematics (Small, 2005a, 2005b, 2006, 2007). Having those maps led to the obvious question: What are they good for?

Clearly, a map of student development is most valuable if it can help teachers adjust instruction to meet the needs of students in their classes who are at different levels of development. To that end, I created a number of professional development courses and many presentations wherein I shared with teachers two strategies that are relatively easy to implement and that allow them to make such curriculum adjustments. These strategies are open questions and parallel tasks. Many teachers who have participated in these sessions have reported to me that they found the approaches productive and manageable in the classroom environment. With their encouragement, and at their suggestion, I have written this book to provide even more models for how the strategies can be used.

ORGANIZATION OF THE BOOK

An introductory chapter describes the rationale for differentiating math instruction and explains the two principal strategies that are employed throughout the book: open questions and parallel tasks. Five content chapters then illustrate application of these strategies, followed by a final summary chapter, an appendix containing a template for teachers wishing to develop their own materials, a glossary, a bibliography, and an index.

Chapters 2–6 focus on the five content strands enunciated in the *Principles and Standards for School Mathematics* of the National Council of Teachers of Mathematics (NCTM):

- Number and Operations
- Algebra
- Geometry
- Measurement
- Data Analysis and Probability (NCTM, 2000)

The content strands are not developed sequentially, and Chapters 2–6 can be approached in any order.

Mathematical concepts are addressed in a framework of big ideas, which have been developed by marrying the NCTM process standards of problem solving, reasoning and proof, communicating, connecting, and representing (NCTM, 2000) with the NCTM content standards for the five content strands listed above. Big ideas are statements of fundamental principles and are broadly applicable to multiple grade bands and different developmental levels of understanding.

Within each of the content chapters, the suggested differentiating questions and tasks are divided according to the primary, elementary, and middle-level grade bands set out in the NCTM (2000) standards:

- Prekindergarten–grade 2
- Grades 3–5
- Grades 6–8

The object of differentiation is to teach the same concepts—in this approach, the same big ideas—to students at different developmental levels. The multilayered organization of material within Chapters 2–6 is intended to illustrate broad principles of application of open questions and parallel tasks while providing ready access to specific concepts at the three grade band levels.

The Appendix features a template worksheet that will assist teachers in developing their own materials in support of differentiated instruction through use of open questions and parallel tasks. An example of application of the worksheet appears in Chapter 1.

The Glossary defines technical terms used throughout. Each word that appears in the Glossary is shown in boldface type at its first occurrence in the text, and each Glossary entry is annotated with the chapter and page number of the term's first occurrence.

The Bibliography highlights three types of resources: those that are referenced in other parts of the text, those that speak to the issues of teaching to big ideas and differentiating instruction, as well as a number of excellent sources for activities that can be used as is or used as a starting point for creating open questions and parallel tasks.

The Index focuses on educational concepts—standards, student development, teaching methods and principles, and such—as opposed to mathematical concepts. To facilitate user access to the mathematical topics covered, an Index of Big Ideas is provided, listing all big ideas covered in the content chapters.

ORGANIZATION OF THE CONTENT CHAPTERS

Chapters 2–6 address the five NCTM content strands, providing examples of open questions and parallel tasks—organized around big ideas—for the prekindergarten–grade 2, grades 3–5, and grades 6–8 grade bands.

Each chapter begins with a listing of the goals of the NCTM standard for the particular content strand, followed by a brief description of how student understanding of the content develops across grade levels from prekindergarten through grade 8. For each grade band, concepts students will learn and apply are described, demonstrating how basic concepts form the foundation for more complex concepts as understanding develops. The content standards are approached through exploration of big ideas, which are listed at the beginning of each chapter.

The bulk of each chapter is composed of a series of illustrations of application of the two differentiation strategies that are the focus of this book: open questions and parallel tasks. Each of these strategies is discussed for each of the three grade bands. Within each grade band section, content is organized by big idea, often with several illustrations for each big idea. Readers may choose to focus on the grade band in which they teach, or they may choose to study all of the grade bands to get a better sense of how understanding of the content develops in students.

For many of the questions and tasks described, important follow-up questions that can be used in the classroom are listed. In many cases, variations of activities or follow-up questions are provided to allow for even more flexibility in differentiating instruction. In addition, the rationale for each activity is presented as background that will make it easier for teachers to develop their own related activities.

Numerous Teaching Tips are included in each chapter. These sometimes relate to a specific activity, but often they are general strategies that can apply to any situation in which a teacher is attempting to differentiate instruction in math.

At the end of each chapter, concluding remarks include a few suggestions for developing additional open questions and parallel tasks by using the template provided in the Appendix.

IT IS MY HOPE that teachers will embrace the two core strategies—open questions and parallel tasks—that are explained and demonstrated in this book, and find them, as I have, to be helpful to the many children who come into classrooms with highly differentiated mathematical preparation, skill, and confidence. Seeing a child who has been struggling in mathematics start to feel successful is an important goal in my teaching. I have seen the use of the strategies described in this volume make that happen over and over again.

Acknowledgments

THIS BOOK rose out of extraordinary opportunities provided to me to conduct research and follow it up with the creation of professional development courses and presentations. These experiences have had significant impact on the direction of my career and have led me to a focus on differentiating instruction in math.

In particular, I thank PRIME project managers Jan Elliott, Rob Greenaway, Deborah Millard, Audrey Wearn, and Lenore Brooks. I also thank the editors with whom I worked on projects that led to this volume, particularly Jackie Williams, as well as a colleague, Doug Duff, who helped me fine-tune the materials I created.

This book is also a direct result of opportunities provided to me by the National Council of Supervisors of Mathematics and the National Council of Teachers of Mathematics to present these ideas at their conferences. Jean Ward, having seen one of those presentations, invited me to work on this book. I thank her for her confidence in me and her assistance in making preparation of the volume as painless as it was.

Why and How to Differentiate Math Instruction

STUDENTS IN ANY CLASSROOM differ in many ways, only some of which the teacher can reasonably attend to in developing instructional plans. Some differences will be cognitive—for example, what previous concepts and skills students can call upon. Some will be more about learning style and preferences, including behaviors such as persistence or inquisitiveness or the lack thereof; whether the student learns better through auditory, visual, or kinesthetic approaches; and personal interests.

THE CHALLENGE IN MATH CLASSROOMS

Although many teachers of language arts recognize that different students need different reading material, depending on their reading level, it is much less likely that teachers vary the material they ask their students to work with in mathematics. The math teacher will more frequently teach all students based on a fairly narrow curriculum goal presented in a textbook. The teacher will recognize that some students need additional help and will provide as much support as possible to those students while the other students are working independently. Perhaps this occurs because **differentiating instruction** in mathematics is a relatively new idea. Perhaps it is because it is not easy in mathematics to simply provide an alternate book to read (as can be done in language arts). And perhaps it is because teachers may never have been trained to really understand how students differ mathematically. However, students in the same grade level clearly *do* differ mathematically in significant ways. Teachers want to be successful in their instruction of all students, and under new laws they are mandated to do so. Understanding differences and differentiating instruction are important processes for achievement of that goal.

The National Council of Teachers of Mathematics (NCTM), the professional organization whose mission it is to promote, articulate, and support the best possible teaching and learning in mathematics, recognizes the need for differentiation. The first principle of the NCTM *Principles and Standards for School Mathematics* reads, "Excellence in mathematics education requires equity—high expectations and strong support for all students" (NCTM, 2000, p. 12).

In particular, NCTM recognizes the need for accommodating differences among students, taking into account prior knowledge and intellectual strengths, to ensure that each student can learn important mathematics. "Equity does not mean that every student should receive identical instruction; instead, it demands that reasonable and appropriate accommodations be made as needed to promote access and attainment for all students" (NCTM, 2000, p. 12).

How Students Might Differ

One way that we see the differences in students is through their responses to the mathematical questions and problems that are put to them. For example, consider the task below, which might be asked of 3rd-grade students:

> In one cupboard, you have three shelves with five boxes on each shelf. There are three of those cupboards in the room. How many boxes are stored in all three cupboards?

Students might approach the task in very different ways. Here are some examples:

- Liam immediately raises his hand and simply waits for the teacher to help him.
- Angelita draws a picture of the cupboards, the shelves, and the boxes and counts each box.
- Tara uses addition and writes $5 + 5 + 5 + 5 + 5 + 5 + 5 + 5 + 5$.
- Dejohn uses addition and writes $5 + 5 + 5 = 15$, then adds again, writing $15 + 15 + 15 = 45$.
- Rebecca uses a combination of multiplication and addition and writes $3 \times 5 = 15$, then $15 + 15 + 15 = 45$.

The Teacher's Response

What do all these different student approaches mean for the teacher? They demonstrate that quite different forms of feedback from the teacher are needed to support the individual students. For example, the teacher might wish to:

- Follow up with Tara and Dejohn by introducing the benefits of using a multiplication expression to record their thinking.
- Help Rebecca extend what she already knows about multiplication to more situations.
- Encourage Liam to be more independent, or set out a problem that is more suitable to his developmental level.
- Open Angelita up to the value of using more sophisticated strategies by setting out a problem in which counting becomes even more cumbersome.

These differences in student approaches and appropriate feedback underscore the need for a teacher to know where his or her students are developmentally to be able to meet each one's educational needs. The goal is to remove barriers to learning while still challenging each student to take risks and responsibility for learning (Karp & Howell, 2004).

WHAT IT MEANS TO MEET STUDENT NEEDS

One approach to meeting each student's needs is to provide tasks within each student's **zone of proximal development** and to ensure that each student in the class has the opportunity to make a meaningful contribution to the class community of learners. Zone of proximal development is a term used to describe the "distance between the actual development level as determined by independent problem solving and the level of potential development as determined through problem solving under adult guidance or in collaboration with more capable peers" (Vygotsky, 1978, p. 86).

Instruction within the zone of proximal development allows students, whether through guidance from the teacher or through working with other students, to access new ideas that are close enough to what they already know to make the access feasible. Teachers are not using educational time optimally if they either are teaching beyond a student's zone of proximal development or are providing instruction on material the student already can handle independently. Although other students in the classroom may be progressing, the student operating outside his or her zone of proximal development is often not benefiting from the instruction.

For example, a teacher might be planning a lesson on multiplying a decimal by a whole number. Although the skill that is the goal of the lesson is to perform a computation such as 3×1.5, there are three underlying mathematical concepts that a teacher would want to ensure that students understand. Students working on this question should know:

- What multiplication means (whether repeated addition, or the counting of equal groups, or calculating the area of a rectangle)
- That multiplication has those same meanings regardless of what number 3 is multiplying
- That multiplication can be accomplished in parts (the distributive principle), for example, $3 \times 1.5 = 3 \times 1 + 3 \times 0.5$

Although the planned lesson is likely to depend on the fact that students understand that 1.5 is 15 tenths or 1 and 5 tenths, a teacher could effectively teach the same lesson even to students who do not have that understanding or who simply are not ready to deal with decimals. The teacher could allow the less developed students to explore the concepts using whole numbers while the more advanced students are using decimals. Only when the teacher felt that the use of decimals was in an individual student's zone of proximal development would the teacher ask that student to work with decimals. Thus, by making this adjustment, the teacher differentiates the task to locate it within each student's zone of proximal development.

ASSESSING STUDENTS' NEEDS

For a teacher to teach to a student's zone of proximal development, first the teacher must determine what that zone is by gathering diagnostic information to assess the student's mathematical developmental level. For example, to determine a 3rd- or 4th-grade student's developmental level in multiplication, a teacher might use a set of questions to find out whether the student knows various meanings of multiplication, knows to which situations multiplication applies, can solve simple problems involving multiplication, and can multiply single-digit numbers, using either memorized facts or strategies that relate known facts to unknown facts (e.g., knowing that 6×7 must be 7 more than 5×7).

Some tools to accomplish this sort of evaluation are tied to developmental continua that have been established to describe students' mathematical growth (Small, 2005a, 2005b, 2006, 2007). Teachers might also use locally or personally developed diagnostic tools. Only after a teacher has determined a student's level of mathematical sophistication, can he or she even begin to attempt to address that student's needs.

PRINCIPLES AND APPROACHES TO DIFFERENTIATING INSTRUCTION

Differentiating instruction is not a new idea, but the issue has been gaining an ever higher profile for mathematics teachers in recent years. More and more, educational systems and parents are expecting the teacher to be aware of what each individual student needs and to plan instruction to focus on those needs. In the past, this was less the case in mathematics than in other subject areas, but now the expectation is common in mathematics as well.

There is general agreement that to effectively differentiate instruction, the following elements are needed:

- *Big Ideas.* The focus of instruction must be on the **big ideas** being taught to ensure that they all are addressed, no matter at what level.
- *Choice.* There must be some aspect of choice for the student, whether in content, process, or product.
- *Preassessment.* Prior assessment is essential to determine what needs different students have (Gregory & Chapman, 2006; Murray & Jorgensen, 2007).

Teaching to Big Ideas

The Curriculum Principle of the NCTM *Principles and Standards for School Mathematics* states that "A curriculum is more than a collection of activities: it must be coherent, focused on important mathematics, and well articulated across the grades" (NCTM, 2000, p. 14).

Curriculum coherence requires a focus on interconnections, or big ideas. Big ideas represent fundamental principles; they are the ideas that link the specifics. For example, the notion that **benchmark numbers** are a way to make sense of other numbers is equally useful for the 1st-grader who relates the number 8 to the more familiar 10, the 3rd-grader who relates 93 to the more familiar 100, or the 8th-grader who relates π to the number 3. If students in a classroom differ in their readiness, it is usually in terms of the specifics and not the big ideas. Although some students in a classroom where rounding of decimal thousandths to appropriate benchmarks is being taught might not be ready for that precise topic, they could still deal with the concept of estimating, when it is appropriate, and why it is useful.

Big ideas can form a framework for thinking about "important mathematics" and supporting standards-driven instruction. Big ideas find application across all grade bands. There may be differences in the complexity of their application, but the big ideas remain constant. Many teachers believe that curriculum requirements limit them to fairly narrow learning goals and feel that they must focus instruction on meeting those specific student outcomes. Differentiation requires a different approach, one that is facilitated by teaching to the big ideas.

Choice

Few math teachers are comfortable with the notion of student choice except in the rarest of circumstances. They worry that students will not make "appropriate" choices.

However, some teachers who are uncomfortable differentiating instruction in terms of the main lesson goal are willing to provide some choice in follow-up activities students use to practice the ideas they have been taught. Some of the strategies that have been suggested for differentiating practice include use of menus from which students choose from an array of tasks, tiered lessons in which teachers teach to the whole group and vary the follow-up for different students, learning stations where different students attempt different tasks, or other approaches that allow for student choice, usually in pursuit of the same basic overall lesson goal (Tomlinson, 1999; Westphal, 2007).

For example, a teacher might present a lesson on creating equivalent fractions to all students, and then vary the follow-up. Some students might work only with simple fractions and at a very concrete level; these tasks are likely to start with simple fractions where the numerator and denominator have been multiplied (but not divided) to create equivalent fractions. Other students might be asked to work at a pictorial or even symbolic level with a broader range of fractions, where numerators and denominators might be multiplied or divided to create equivalent fractions and more challenging questions are asked (e.g., *Is there an equivalent fraction for $\frac{10}{15}$ where the denominator is 48?*).

Preassessment

To provide good choices, a teacher must first know how students in the classroom vary in their knowledge of facts and in their mathematical developmental level. This requires collecting data either formally or informally to determine what abilities and what deficiencies students have. Although many teachers feel they lack the time or the tools to preassess on a regular basis, the data derived from preassessment are essential in driving differentiated instruction.

Despite the importance of preassessment, employing a highly structured approach or a standardized tool for conducting the assessment is not mandatory. Depending on the topic, a teacher might use a combination of written and oral questions and tasks to determine an appropriate starting point for each student.

TWO CORE STRATEGIES FOR DIFFERENTIATING MATHEMATICS INSTRUCTION: OPEN QUESTIONS AND PARALLEL TASKS

It is not realistic for a teacher to try to create 30 different instructional paths for 30 students, or even 6 different paths for 6 groups of students. Because this is the perceived alternative to one-size-fits-all teaching, instruction in mathematics is often not differentiated. To differentiate instruction efficiently, teachers need manageable strategies that meet the needs of most of their students at the same time. Through use of just two core strategies, teachers can effectively differentiate instruction to suit all students. These two core strategies are the central feature of this book:

- **Open questions**
- **Parallel tasks**

Open Questions

The ultimate goal of differentiation is to meet the needs of the varied students in a classroom during instruction. This becomes manageable if the teacher can create a single question or task that is inclusive not only in allowing for different students to approach it by using different processes or strategies but also in allowing for students at different stages of mathematical development to benefit and grow from attention to the task. In other words, the task is in the appropriate zone of proximal development for the entire class. In this way, each student becomes part of the larger learning conversation, an important and valued member of the learning community. Struggling students are less likely to be the passive learners they so often are (Lovin, Kyger, & Allsopp, 2004).

A question is open when it is framed in such a way that a variety of responses or approaches are possible. Consider, for example, these two questions, each of which might be asked of a whole class, and think about how the results for each question would differ:

Question 1: To which fact family does the fact $3 \times 4 = 12$ belong?

Question 2: Describe the picture below by using a mathematical equation.

x	x	x	x
x	x	x	x
x	x	x	x

If the student does not know what a fact family is, there is no chance he or she will answer Question 1 correctly. In the case of Question 2, even if the student is not comfortable with multiplication, the question can be answered by using addition statements (e.g., $4 + 4 + 4 = 12$ or $4 + 8 = 12$). Other students might use multiplication statements (e.g., $3 \times 4 = 12$ or $4 \times 3 = 12$), division statements (e.g., $12 \div 3 = 4$ or $12 \div 4 = 3$), or even statements that combine operations (e.g., $3 \times 2 + 3 \times 2 = 12$).

A Different Kind of Classroom Conversation. Not only will the mathematical conversation be richer in the case of Question 2 on the previous page—the open question—but almost any student can find something appropriate to contribute.

The important point to notice is that the teacher can put the same question to the entire class, but the question is designed to allow for differentiation of response based on each student's understanding. All students can participate fully and gain from the discussion in the classroom learning community.

This approach differs, in an important way, from asking a question, observing students who do not understand, and then asking a simpler question to which they can respond. By using the open question, students gain confidence; they can answer the teacher's question right from the start. Psychologically, this is a much more positive situation.

Multiple Benefits. There is another benefit to open questions. Many students and many adults view mathematics as a difficult, unwelcoming subject because they see it as black and white. Unlike, for instance, social studies or English, where people might be encouraged to express different points of view, math is viewed as a subject where either you get it or you don't. This view of mathematics inhibits many students from even trying. Once they falter, they assume they will continue to falter and may simply shut down.

It is the job of teachers to help students see that mathematics is multifaceted. Any mathematical concept can be considered from a variety of perspectives, and those multiple perspectives actually enrich its study. Open questions provide the opportunity to demonstrate this.

Strategies for Creating Open Questions. This book illustrates a variety of styles of open questions. Some common strategies that can be used to convert conventional questions to open questions are described below:

- Turning around a question
- Asking for similarities and differences
- Replacing a number with a blank
- Asking for a number sentence
- Changing the question

Turning Around a Question. For the turn-around strategy, instead of giving the question, the teacher gives the answer and asks for the question. For example:

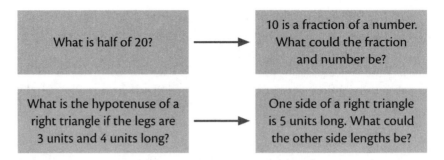

Asking for Similarities and Differences. The teacher chooses two items—two numbers, two shapes, two graphs, two probabilities, two measurements, and so forth—and asks students how they are alike and how they are different. Inevitably, there will be many good answers. For example, the teacher could ask how the number 85 is like the number 100 and how it is different. A student might realize that both numbers are said when skip counting by 5s, both are less than 200, and both are greater than 50, but only one is a three-digit number, only one ends with a 5, and only one is greater than 90.

Replacing a Number with a Blank. Open questions can be created by replacing a number with a blank and allowing the students to choose the number to use. For example, instead of asking how many students there are altogether if there are 25 in one class and 31 in another, students could be asked to choose two numbers for the two class sizes and determine the total number in both classes.

Asking for a Number Sentence. Students can be asked to create a sentence that includes certain words and numbers. For example, a teacher could ask students to create a sentence that includes the numbers 3 and 4 along with the words "and" and "more," or a sentence that includes the numbers 8 and 7 as well as the words "product" and "equal." The variety of sentences students come up will often surprise teachers. In the first case, students might produce any of the sentences below and many more:

- <u>3 *and* 4</u> are <u>*more*</u> than 2.
- <u>4</u> is <u>*more*</u> than <u>3 *and* *more*</u> than 1.
- <u>3 *and* 4</u> together are <u>*more*</u> than 6.
- <u>34 *and* 26</u> are <u>*more*</u> than <u>34 *and*</u> 20.

Changing the Question. A teacher can sometimes create an open question by beginning with a question already available, such as a question from a text resource. Here are a few examples:

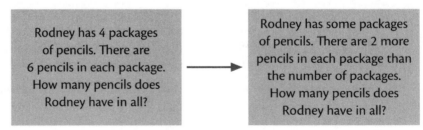

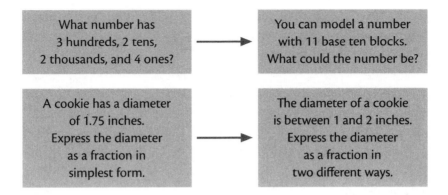

What to Avoid in an Open Question. An open question should be mathematically meaningful. There is nothing wrong with an occasional question such as *What does the number 6 make you think of?* but questions that are focused more directly on big ideas or on curricular goals are likely to accomplish more in terms of helping students progress satisfactorily in math.

Open questions need just the right amount of ambiguity. They may seem vague, and that may initially bother students, but the vagueness is critical to ensuring that the question is broad enough to meet the needs of all students.

On the other hand, one must be careful about making questions so vague that they deter thinking. Compare, for example, a question like *What's a giant step?* with a question like *How many baby steps are in a giant step?* In the first case, a student does not know whether what is desired is a definition for the term, a distance, or something else. The student will most likely be uncomfortable proceeding without further direction. In the second case, there is still ambiguity. Some students may wonder what is meant by "baby step" and "giant step," but many will be comfortable proceeding; they realize they are being asked about a measurement situation.

The reason for a little ambiguity is to allow for the differentiation that is the goal in the use of open questions. Any question that is too specific may target a narrow level of understanding and not allow students who are not at that level to engage with the question and experience success.

Fostering Effective Follow-Up Discussion. Follow-up discussions play a significant role in cementing learning and building confidence in students. Thus, it is important for teachers to employ strategies that will optimize the effectiveness of follow-up discussions to benefit students at all developmental levels.

To build success for all students, it is important to make sure that those who are more likely to have simple answers are called on first. By doing so, the teacher will increase the chances that these students' answers have not been "used up" by the time they are called on.

The teacher must convey the message that a variety of answers are appreciated. It is obvious to students when a teacher is "looking for" a particular answer. An open question is designed to ensure that many answers are good answers and will be equally valued.

The teacher should try to build connections between answers that students provide. For example, when asked how 10 and 12 are alike, one student might say that both numbers are even and another might say that they are both between 10 and 20. The teacher could follow up with:

- *What other even numbers are there between 10 and 20?*
- *Which digits tell you the numbers are even?*
- *Which digits tell you the numbers are between 10 and 20?*

Such questions challenge all students and scaffold students who need help.

Parallel Tasks

Parallel tasks are sets of tasks, usually two or three, that are designed to meet the needs of students at different developmental levels, but that get at the same big idea and are close enough in context that they can be discussed simultaneously. In other words, if a teacher asks the class a question, it is pertinent to each student, no matter which task that student completed. The use of parallel tasks is an extension of Forman's (2003) point that task modification can lead to valuable discussions about the underlying mathematics of a situation. Parallel tasks also contribute to the creation of the classroom as a learning community in which all students are able to contribute to discussion of the topic being studied (Murray & Jorgensen, 2007).

For example, suppose a teacher wishes to elicit the big idea within the NCTM Numbers and Operations strand that it is important to recognize when each mathematical operation is appropriate to use. The teacher can set out two parallel tasks:

> _Option 1:_ Create a word problem that could be solved by multiplying two one-digit numbers.
>
> _Option 2:_ Create a word problem that could be solved by multiplying two numbers between 10 and 100.

Both options focus on the concept of recognizing when multiplication is appropriate, but Option 1 is suitable for students only able to work with smaller factors. Further, the tasks fit well together because questions such as the ones listed below suit a discussion of both tasks and thus can be asked of all students in the class:

- *What numbers did you choose to multiply?*
- *How did you know how many digits the product would have?*
- *What about your problem made it a multiplying problem?*
- *What was your problem?*
- *How could you solve it?*

Strategies for Creating Parallel Tasks. To create parallel tasks to address a particular big idea, it is important to first think about how students might differ developmentally in approaching that idea. Differences might relate to what operations the students can use or what size numbers they can handle, or they might involve, for example, what meanings of an operation make sense to the students.

Once the developmental differences have been identified, the object is to develop similar enough contexts for the various options that common questions can be asked of the students as they reflect on their work. For example, for the big idea that standard measures simplify communication, the major developmental difference might be the type of measurement with which students are comfortable. One task could focus on linear measurements and another on area measurements. One set of parallel tasks might be:

> _**Option 1:**_   An object has a length of 30 cm. What might it be?
>
> _**Option 2:**_   An object has an area of 30 cm². What might it be?

In this case, common follow-up questions could be:

- _Is your object really big or not so big? How did you know?_
- _Could you hold it in your hand?_
- _How do you know that your object has a measure of about 30?_
- _How would you measure to see how close to 30 it might be?_
- _How do you know that there are a lot of possible objects?_

Often, to create a set of parallel tasks, a teacher can select a task from a handy resource (e.g., a student text) and then figure out how to alter it to make it suitable for a different developmental level. Then both tasks are offered simultaneously as options for students:

> _**Original task**_ (e.g., from a text):
> There were 483 students in the school in the morning. 99 students left for a field trip. How many students are left in the school?
>
> _**Parallel task:**_
> There are 71 students in 3rd grade in the school. 29 of them are in the library. How many are left in their classrooms?

Common follow-up questions could be:

- _How do you know that most of the students were left?_
- _How did you decide how many were left?_
- _Why might someone subtract to answer the question?_
- _Why might someone add to answer the question?_
- _How would your answer have changed if one more student had left?_
- _How would your answer have changed if there had been one extra student to start with?_

Fostering Effective Follow-Up Discussion. The role of follow-up discussions of parallel tasks and the techniques for encouraging them mirror those for open questions. Once again, it is critical that the teacher demonstrate to students that he or she values the tasks equally by setting them up so that common questions suit each of them. It is important to make sure students realize that the teacher is equally interested in responses from the groups of students pursuing each of the options. The teacher should try not to call first on students who have completed one of the tasks and then on students who have completed the other(s). Each question should be addressed to the whole group. If students choose to identify the task they selected, they may, but it is better if the teacher does not ask which task was performed as the students begin to talk.

Management Issues in Choice Situations. Some teachers are concerned that if tasks are provided at two levels, students might select the "wrong" task. It may indeed be appropriate at times to suggest which task a student might complete. This could be done by simply assigning a particular task to each student. However, it is important sometimes—even most of the time—to allow the students to choose. Choice is very empowering.

If students who struggle with a concept happen to select a task beyond their ability, they will soon realize it and try the other option. Knowing that they have the choice of task should alleviate any frustration students might feel if they struggle initially. However, students may also sometimes be able to complete a task more challenging than they first thought they could handle. This would be a very positive experience.

If students repeatedly select an easier task than they are capable of, they should simply be allowed to complete the selected task. Then, when they are done, the teacher can encourage them privately to try the other option as well.

Putting Theory into Practice

A form such as the one that appears on the next page can serve as a convenient template for creation of customized materials to support differentiation of instruction in math. In this example, a teacher has developed a plan for differentiated instruction on the topic of measurement. A blank form is provided in the Appendix.

The following fundamental principles should be kept in mind when developing new questions and tasks:

- All open questions must allow for correct responses at a variety of levels.
- Parallel tasks need to be created with variations that allow struggling students to be successful and proficient students to be challenged.
- Questions and tasks should be constructed in such a way that all students can participate together in follow-up discussions.

Teachers may find it challenging at first to incorporate the core strategies of open questions and parallel tasks into their teaching routines. However, after trying

MY OWN QUESTIONS AND TASKS

Lesson Goal: Area measurement   **Grade Level:** 4

Standard(s) Addressed:

Choose appropriate customary and metric units to estimate and measure length, perimeter, area, weight, capacity, volume, and temperature, including square feet and square inches . . .

Underlying Big Idea(s):

The same object can be described by using different measurements.

Open Question(s):

Which shape is bigger? How do you know?

Parallel Tasks:

Option 1:

Which shape has a greater perimeter? How much greater is it?

Option 2:

Which shape has a greater area? How much greater is it?

Principles to Keep in Mind:

- All open questions must allow for correct responses at a variety of levels.
- Parallel tasks need to be created with variations that allow struggling students to be successful and proficient students to be challenged.
- Questions and tasks should be constructed in such a way that will allow all students to participate together in follow-up discussions.

examples found in the five chapters that follow, and creating their own questions and tasks, teachers will soon find that these strategies become second nature. And the payoff for the effort will be the very positive effects of differentiation that emerge: fuller participation by all students and greater advancement in learning for all.

CREATING A MATH TALK COMMUNITY

Throughout this book, many suggestions will be offered for ways to differentiate instruction. These are all predicated on a classroom climate where mathematical conversation is the norm, a variety of student approaches are encouraged and valued, and students feel free to take risks. Unless a student engages in mathematical conversation, it is not possible for a teacher to know what that individual does or does not understand. Consequently, it is not possible for a teacher to know what steps must be taken to ensure the student's needs are being met.

Some teachers may be nervous about offering choices or asking open questions—worried that students might get off track, worried that students might be uncomfortable with not being absolutely certain about what is expected, or worried that students might offer an idea with which the teacher is unfamiliar, leaving the teacher unsure about how to proceed. These are natural concerns.

Initially, students who are accustomed to highly structured learning environments may find open questions or choice unsettling. But once the students see the teacher's willingness to allow them to go in different directions, they will grow comfortable with the change and will appreciate the opportunity for greater input. Teachers will also find it both surprising and rewarding to see how students rise to the challenge of engaging in mathematical conversation and how students often help the teacher sort out an unclear comment from another student or suggest ways to pick up on another student's suggestion.

Number and Operations

DIFFERENTIATED LEARNING activities in number and operations are derived from applying the NCTM process standards of problem solving, reasoning and proof, communicating, connecting, and representing to content goals of the NCTM Number and Operations Standard, including

- understanding numbers, ways of representing numbers, relationships among numbers, and number systems
- understanding meanings of operations and how they relate to one another
- computing fluently and making reasonable estimates (NCTM, 2000)

TOPICS

Before differentiating instruction in number and operations, it is useful for a teacher to have a sense of what number concepts students typically bring to a grade level and how concepts in number and operations tend to develop after that level. The NCTM Curriculum Focal Points (NCTM, 2006) are one source of this knowledge, as are state or local standards and research findings.

Students move from working comfortably with relatively small numbers concretely to working with whole numbers up to 1,000 symbolically to working with fractions, decimals, and larger whole numbers (Small, 2005a). They move from solving problems involving addition, subtraction, multiplication, and division by counting to solving problems using strategies and learned and invented procedures; as students develop mathematically, they flexibly use more efficient procedures and strategies.

Prekindergarten–Grade 2

Within this grade band, students move from counting and comparing very simple numbers, usually 10 or less, to counting, comparing, modeling, and interpreting numbers up to 1,000 using place value concepts. They move from counting to determine sums in simple joining situations and differences in simple subtraction situations to thinking more formally about adding and subtracting. Increasingly, as they move through the grade band, students use a variety of principles, strategies, and procedures with increasing efficiency to add and subtract and to solve problems requiring addition and subtraction.

Grades 3–5

Within this grade band, students begin to focus increasingly on multiplying and dividing whole numbers using a variety of strategies to calculate and estimate **products** and then **quotients**. They commit multiplication and related division facts to memory, become more fluent with **algorithms** for multiplying and dividing multidigit whole numbers, and solve problems that represent a variety of meanings of multiplication and division.

Students also begin to develop a greater understanding of fractions, first modeling, representing, and comparing them, and later using **equivalent fractions** to simplify the task and to add and subtract fractions. They start to use decimal notation to represent numbers that have fractional parts and learn to add and subtract decimals to solve problems that require these operations.

Grades 6–8

Within this grade band, students extend their understanding of fractions and decimals to situations involving multiplication and division of these values. They increasingly work with ratio and **proportion**, particularly, but not exclusively, in percentage situations. They also begin to work with more abstract values, including **negative integers**, **exponents**, and **scientific notation**.

THE BIG IDEAS FOR NUMBER AND OPERATIONS

Coherent curricula in number and operations that meet NCTM content and process standards (NCTM, 2000) and support differentiated instruction can be structured around the following big ideas:

- There are many ways to represent numbers.
- Numbers tell how many or how much.
- Number benchmarks (referent numbers that are familiar and meaningful, such as 10, 25, 100, 1,000, etc.) are useful for relating numbers and estimating amounts.
- By classifying numbers (e.g., in terms of how many digits they have, whether they are odd or even, etc.), conclusions can be drawn about them.
- The patterns in the **place value system** can make it easier to interpret and operate with numbers.
- It is important to recognize when each operation (addition, subtraction, multiplication, or division) is appropriate to use.
- There are many different ways to add, subtract, multiply, or divide numbers.
- It is important to use and take advantage of the relationships between the operations in computational situations.

The tasks set out and the questions asked while teaching number and operations should be developed to reinforce these ideas. The following sections present numerous examples of application of open questions and parallel tasks in development of differentiated instruction in these big ideas across three grade bands.

OPEN QUESTIONS FOR PREKINDERGARTEN–GRADE 2

OPEN QUESTIONS are broad-based questions that invite meaningful responses from students at many developmental levels.

✴ **BIG IDEA.** **There are many ways to represent numbers.**

> What makes 5 a special number?

Posing this question to young students provides them an opportunity to participate in a mathematical conversation. Some students might think of the fact that there are 5 fingers on a hand, others that they are 5 years old, others that there are 5 people in their family, whereas others might think of something else, for example, that there is a special coin for 5 cents. The question helps students recognize that numbers are used to describe amounts in a wide variety of contexts.

Variations. This question can be varied by using other numbers that students might view as special, for example, 0, 1, or 10.

> Show the number 7 in as many different ways as you can.

To respond to this question, some students will draw seven similar items, others will draw seven random items, others will use stylized numerals, others will write the word *seven,* and still others will show mathematical operations that produce an answer of 7, such as 6 + 1 or 4 + 3.

Because the question asks students to show as many representations as they can, even a student who comes up with only one or two representations will experience success. No matter what their development level, all students will focus on the fact that any number can be represented in many ways.

TEACHING TIP. Many questions can be varied very simply for use over and over again. The question here is just one example. Although the number 7 was selected in this case, the question can be opened up even more by asking students to select a number of their own and show it in as many ways as they can.

> Choose two of these numerals. How do they look the same? How do they look different?
>
> 1   2   3   4   5   6   7   8   9   0

An important part of mathematical development is the recognition that numbers can be used to represent quantities, but it helps if students recognize the numerals. By focusing on the form of the numerals, this question helps students learn to reproduce and read them.

By allowing students to choose whatever two numerals they wish, the question allows for virtually every student to succeed. For example, a very basic comparison might be that 1 and 4 both use only straight lines, but fewer lines are needed to form a 1 than to form a 4.

> The answer is 5. What is the question?

This question has innumerable possible answers, and it supports the important concept that numbers can be represented in many ways. Some students will use addition and ask the question: *What is 4 + 1?* Other students will use subtraction and ask: *What is 6 − 1?* Yet other students will ask very different types of questions, for example: *What number comes after 4?* or even *How many toes are there on one foot?* Because of the wide range of acceptable responses, students at all levels are addressed.

Variations. This task can be reassigned after changing the number required for the answer.

✹ **BIG IDEA.** **Numbers tell how many or how much.**

> Tell about a time when you would use the number $\frac{1}{2}$.

All students who address this question will think about the fraction $\frac{1}{2}$. Allowing them to come up with their own contexts lets each one enter the mathematical conversation at an appropriate level. Often, when a teacher preselects a context for students, it is meaningful to only some of them.

If students need a stimulus to get started, the teacher can provide drawings that might evoke ideas, for example, a picture showing a sandwich cut in two, a glass half full of juice, or a hexagonal table with a line dividing it in half and two people sitting at opposite sides of the table. It is likely that most students will only think of halves of wholes, but some might think of half of a set of two.

TEACHING TIP. Scaffolding tasks by providing models for students reduces the likely breadth of responses that will be received and may inhibit some students who feel obligated to follow the models.

Ellen ran in a race with some other students. She did not win, but she was not last.
- Draw a picture that shows how she did in the race.
- Tell or show on the picture how many people were ahead of her.
- Tell or show on the picture how many people were behind her.
- What word describes her position?

This task allows students to choose comfortable numbers of racers with which to work. No matter what numbers students choose as the number of racers ahead of Ellen and the number of racers behind her, they have the opportunity to show an understanding of both **ordinal numbers** and **cardinal numbers**. If students are familiar with addition, a final question could be added to ask them to tell what addition equation their picture shows.

For example, a child who draws a picture with four students, with Ellen in the second position, would observe that one person was ahead of Ellen, two were behind her, and Ellen was second. The child might also record the addition sentence $1 + 1 + 2 = 4$. Another student might draw a picture with nine students, with Ellen in the middle. That student would observe that four people were ahead of Ellen, four were behind her, and Ellen was fifth. That child's addition sentence would be $4 + 1 + 4 = 9$.

As students hear the variety of responses, they will observe some similarities in the answers, particularly that the number in the middle of one side of the addition equation is 1.

To challenge students who have a more sophisticated understanding of number, an additional condition can be imposed, for example, that there were the same number of runners ahead of Ellen as behind her, that the total number of racers was 32, or that there were more than 15 racers altogether.

✳ **BIG IDEA.** **Number benchmarks are useful for relating numbers and estimating amounts.**

> Choose a number for the second mark on the **number line**.
>
> |———————————————————|————————————▶
> 0
>
> Mark a third point on the line. Tell what number name it should have and why.

This question helps students see the value of using benchmarks to relate numbers. It is designed to be open by allowing students not only to choose which point to locate but also which benchmark to use.

A simple response might be to designate the second marked point as a 2 and to mark a halfway point as a 1. A more complex response might be to designate the second marked point as a 10 and attempt to locate a number like 2 or 3.

No matter which choice students make, they can participate in the conversation about how numbers are placed on a number line and how numbers relate to one another.

Variations. The teacher can name the second point, for example as 5, then ask students to choose the third point.

> Create a sentence that uses each of the four numbers and words shown below. Other words or numbers can also be used.
>
> *3, more, 5, and*

Answers to this question can vary a great deal. Students of different developmental levels might write:

- 3 <u>and</u> 5 are <u>more</u> than 2.
- 5 is <u>more</u> than 3, <u>and</u> 2 is more than 1.
- 53 + 3 is <u>more</u> than 28 <u>and</u> 10.

Although some answers are much simpler than others, all are useful contributions to a discussion of number comparison.

✳ **BIG IDEA.** **By classifying numbers, conclusions can be drawn about them.**

> How are the numbers 10 and 15 alike? How are they different?

A question like this one provides the opportunity to see what students know about numbers less than 20. They might observe similarities such as these:

- *They are both less than 20.*
- *They are both more than 5.*
- *You can show both with nickels.*
- *You say both numbers when you count by 5s.*
- *They can both be represented by full rows in 10-frames.*

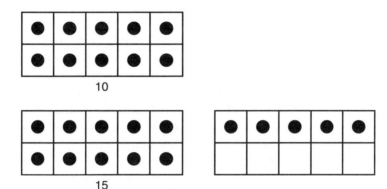

10

15

The students might observe differences such as these:

- *10 is 1 away from 9, but 15 is a lot farther away from 9.*
- *10 is often shown in counting books, but 15 is not.*
- *10 can be shown using one 10-frame, but 15 cannot.*

The question is suitable for a broad range of students because some students can choose simple similarities and differences, such as indicating that both numbers start with a 1, whereas others can focus on more complex similarities, such as indicating that both numbers are part of the sequence when skip counting by 5. All students benefit from the larger discussion about number representations and meanings.

Variations. The question can easily be varied by using other pairs of numbers.

✴ **BIG IDEA.** **The patterns in the place value system can make it easier to interpret and operate with numbers.**

> A two-digit number has more tens than ones. What could the number be? How do you know your number is correct?

Much of the work with numbers is built on the fact that numbers are written in such a way that the value of a numeral is dependent on its placement in the number. For example, the 2 in 23 is worth 20, but the 2 in 32 is worth only 2.

To help students work with the place value system, instruction often begins with models that show the difference. For example, **base ten blocks** show 20 as two rods, but 2 as two small cubes.

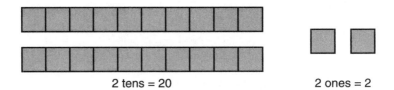

2 tens = 20   2 ones = 2

Students may or may not be provided with base ten blocks when the question above is posed. For most students, provision of the blocks would be helpful, but if students are more advanced, the blocks may not be essential. Some students may discover only one or two possible responses, whereas others might determine many possible responses (e.g., 31, 32, 43, 42, 41, etc.) or even all possible responses. By asking for only one number, the task seems less onerous to struggling students. In discussing responses, students will see that there were many possible values among which they could have chosen.

In showing how they know their number is correct, some students can use concrete models to support their answers, whereas others might use more symbolic arguments.

Variations. Students who are ready to work with three-digit numbers can be given the option to do that if they wish.

✸ BIG IDEA. **There are many different ways to add, subtract, multiply, or divide numbers.**

> Make up an addition question where there is a 2, a 3, and a 4 somewhere in the question or the answer.

All students who respond to the question will use addition, but different students can choose combinations with which they are more comfortable. For example, students might write:

- $2 + \underline{34} = \underline{36}$
- $\underline{2} + \underline{3} + \underline{4} = 9$
- $\underline{23} + \underline{43} + \underline{25} = 91$

An open question such as this promotes a rich discussion. As different individuals or pairs of students share their questions, other students repeatedly have the opportunity to learn new ideas about addition.

Variations. The question can easily be varied by using different sets of three numbers, using two or four numbers, or using a different operation (e.g., subtraction).

Replace the boxes with values from 1 to 6 to make each problem true. You can use each number as often as you want. You cannot use 7, 8, 9, or 0.

This question requires students to add two-digit numbers to a one-digit number or another two-digit number but gives them a choice about what numbers to use.

Students who have more difficulty with adding may choose low values such as 1 and 2 for both the ones and tens places in the missing addends to make the addition easier (e.g., 21 + 2 = 23, 31 + 2 = 33, 22 + 22 = 44). The challenge for other students may be to determine many or all possible combinations.

All students will have the chance to practice addition. In fact, the students will get a great deal of extra practice if they use combinations that have to be discarded. For example, if a student wants to add 24 + 4, the answer would have an 8, which is not allowed.

Variations. The question can easily be varied by using exclusively two-digit or exclusively one-digit numbers for the second addends. It can also be varied by altering the allowable digits. For example, the only allowable digits might be 1, 2, 6, 7, and 8 (with possible solutions of 71 + 6 = 77, 81 + 7 = 88, 62 + 26 = 88, etc.).

Use a **100 chart**. Choose two numbers to add. Both numbers must be on the top half of the chart. Show how to use the chart to add the two numbers without using pencil or paper.

1	2	3	4	5	6	7	8	9	10
11	12	13	14	15	16	17	18	19	20
21	22	23	24	25	26	27	28	29	30
31	32	33	34	35	36	37	38	39	40
41	42	43	44	45	46	47	48	49	50
51	52	53	54	55	56	57	58	59	60
61	62	63	64	65	66	67	68	69	70
71	72	73	74	75	76	77	78	79	80
81	82	83	84	85	86	87	88	89	90
91	92	93	94	95	96	97	98	99	100

This question promotes students' use of visualization to help them with mental math. They can select simple numbers to add or more complicated ones. For example, students who start at 32 and add 1 realize that it is only necessary to move one space to the right. If the student adds 10, it is only necessary to move one space down. A student who adds something like 29, though, might realize that it is possible to go down three rows and back one space, a much more sophisticated mental math operation.

The question is useful for a variety of student levels because students can avoid adding to a number at the right side of the chart (e.g., adding 59 + 6) and having to go to the next line and can avoid adding complicated numbers if they find the maneuvers too difficult.

OPEN QUESTIONS FOR GRADES 3–5

✹ **BIG IDEA.** **There are many ways to represent numbers.**

> The answer is 42. What is the question?

This question can be answered in many ways. Some students will use multiplication and ask the question: *What is 6 × 7?* Other students will use addition or place value and ask: *What is 40 + 2, or 4 tens and 2 ones?* Yet other students will ask very different types of questions, for example: *How much are 8 nickels and 2 pennies worth?* or *How many students are in the library?* With so many possible answers, students at all levels can achieve success.

Variations. Changing the number allows for additional or alternate mathematical conversations. For example, students might find it easier to come up with real-life descriptions for numbers such as 25 than for a number like 42.

> You add two fractions and the sum is $\frac{9}{10}$. What could the fractions be?

Typically, a teacher provides two fractions and asks students for the sum. If, instead, the sum is given and students are asked for the fractions, a broader range of students can respond successfully.

For example, the objective might be to sum two fractions to get $\frac{9}{10}$. A simple response would be $0 + \frac{9}{10}$. An interesting discussion could ensue as students debate whether 0 does or does not have to be written as $\frac{0}{10}$ to satisfy the instruction or to perform the addition.

Other fairly straightforward solutions are $\frac{8}{10} + \frac{1}{10}$, $\frac{7}{10} + \frac{2}{10}$, and so on. These become more obvious to students if they draw a diagram to represent $\frac{9}{10}$ and simply break up the parts. For example, the diagram on the next page shows how $\frac{9}{10}$ can be represented as $\frac{4}{10} + \frac{5}{10}$:

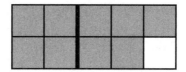

Looking at the representation, some students might even see $\frac{1}{2} + \frac{4}{10}$ if they recognize that the top row is $\frac{1}{2}$ of the whole and shaded boxes in the bottom row represent $\frac{4}{10}$ of the whole.

Yet other students will work symbolically, either looking for fractions with denominators of 10 to add or by writing $\frac{9}{10}$ as an equivalent fraction, for example, $\frac{18}{20}$ or $\frac{27}{30}$, and adding fractions with denominators of 20 or 30, respectively.

Some students are likely to choose only one pair of addends; others may seek many pairs. In this way, a broader range of students is being accommodated. All of the students will recognize that adding is about putting things together.

Variations. It is easy to vary the question by using a different sum or by asking for a particular difference rather than a particular sum.

✳ BIG IDEA. **Numbers tell how many or how much.**

> Why is 0 a special number?

This question focuses on mathematical communication, a skill that is important for all students to develop. Some students may focus on computation—for example, noting that when you add 0 to a number, the number does not change, or that when you multiply a number by 0, the product is 0—whereas other students might come up with more complex ideas.

For example, a student might think about negative numbers and suggest that 0 is the only number that does not have an **opposite** in a different location on the number line. Still others might focus on something unanticipated, for example, suggesting that 0 is the only number that looks like a circle.

All of these are equally interesting observations and all should be equally appreciated.

> A number describes how many students are in a school. What might the number be?

An important part of mathematics instruction is to allow students to make sense of the numbers around them. People need to be able to recognize when a number makes sense and when it does not. For example, students should know that a price of $500 for a pencil makes no sense.

The question above illustrates a real-life context. In this case, the situation is close to students' everyday experience; they can think about how many students are in their own school. However, the question can also lead them to consider other possibilities, for example, how many students there might be in a big high school, a small country school, or a medium-size school.

Some students will use calculations, multiplying the number of students in a class by the number of classes. If students simply guess, they should be expected to justify their values.

> Draw a small rectangle. Draw a bigger rectangle that the smaller one is part of. Tell what fraction of the big rectangle the small one is.

In teaching fractions, students are usually asked to identify a fraction given a partitioned whole. The question above has students thinking the other way—what is the whole if a fractional part is known? Such reversible thinking is an important mathematical process to develop.

To answer a question phrased in this more open way, a struggling student might use a simple fraction such as $\frac{1}{2}$, whereas other students might suggest more complex fractions. Several possibilities are shown in the diagram below:

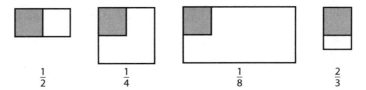

$$\frac{1}{2} \qquad \frac{1}{4} \qquad \frac{1}{8} \qquad \frac{2}{3}$$

✹ **BIG IDEA.** **Number benchmarks are useful for relating numbers and estimating amounts.**

> Choose two numbers to compare. Tell which is greater and how much greater. Tell how you know.

With a choice of numbers to compare, almost any student can simplify the task sufficiently to be successful. For example, one student might choose the numbers 21 and 20, indicating that 21 is 1 greater because it is the next number you would say if you counted. Another student might, instead, choose numbers like 410 and 27, indicating that 410 is greater because it is over 100 and 27 is under 100; he or she might even indicate how much greater 410 is.

TEACHING TIP. One of the simplest strategies for differentiating instruction is allowing students to choose the numbers with which they will work.

> One number is a lot more than another one. Both numbers are greater than 100. What could the two numbers be?

This question provides a great deal of latitude both in the choice of numbers and in the interpretation of the phrase *a lot more*. This ambiguity allows for a richer mathematical discussion.

For example, one student might choose 200 and 300, suggesting that being 100 apart means that 300 is a lot more. Others might argue that those numbers are not far enough apart and might choose, for example, 1,000,000 and 101. Neither student is incorrect, but the door is opened for a discussion about how some mathematical terms are more precise than other terms and how both types of terms can be useful in different situations.

In addition, there is an opportunity within the larger class community to repeatedly practice the concept of comparison.

> You multiply two numbers and the product is almost 400. What could the numbers have been? Explain your answer.

Estimation is an important aspect of mathematical calculation. Teachers often do not emphasize it enough. This open question allows students to work backward and think about what numbers might have been used to arrive at a particular estimate. They will also have to think about what the term *almost* means.

Some students might think of something fairly simple, for example, 1 and 399. Others might recognize that $20 \times 20 = 400$ and then use 19×19. Still others will consider other possibilities.

Variations. Instead of using "almost," the question can be varied by using "a bit more than." The product can be changed from 400 to a different number. Other operations can also be allowed: for example, the question might be worded, "You add, subtract, multiply, or divide two numbers and the result is 400. What could the numbers be? Explain."

✳ BIG IDEA. **By classifying numbers, conclusions can be drawn about them.**

> How are the numbers 350 and 550 alike? How are they different?

Although many students at this level will be comfortable with three-digit numbers, some may still be struggling. Even struggling students can use simple ideas to compare 350 and 550, for example, noticing that both 350 and 550 end in 50 and that they begin with different digits. Other students might use more sophisticated mathematics. For example, they might suggest that both numbers leave no remainder when divided by 10 or 50, that both require more than 5 base ten blocks to model them, or that one is greater than 500 whereas the other is not.

Variations. The question can easily be adapted by using other numbers that are even more obviously similar, for example, 200 and 400.

> Choose two fractions with different **denominators**. Tell how to compare them.

Students can use many different strategies to compare fractions. Some students might choose a fraction less than 1 and a fraction greater than 1 by appropriately manipulating the **numerators** and denominators.

Others might use other benchmarks to make the comparisons simple. For example, they might choose a fraction very close to 0, such as $\frac{1}{1,000}$, and another very close to 1, such as $\frac{99}{100}$. Still others might think of a fraction less than $\frac{1}{2}$, such as $\frac{1}{3}$, and a fraction greater than $\frac{1}{2}$, such as $\frac{3}{4}$.

And other students, perhaps even most students, are likely to select two arbitrary fractions and then use equivalent fractions with a common denominator to compare them.

As students tell how to compare the fractions, they are free to use whatever strategy they wish. This sort of choice is important to allow all students to succeed. It is also helpful to all students, as the comparisons are discussed, to hear about the many different strategies they could have accessed.

Variations. The teacher can require that one or both of the numerators of the two fractions not be 1, because often students tend to use only **unit fractions**.

✷ **BIG IDEA.** The patterns in the place value system can make it easier to interpret and operate with numbers.

> Use 16 base ten blocks to represent a number. What numbers can you represent?

Students should be provided with base ten blocks when this question is posed. Some students might choose to use only ones and tens (i.e., a number such as 88); others might use ones, tens, and hundreds (e.g., 169); still others might use thousands as well (e.g., 3,427).

Typically, a number is named and students are asked to represent it with base ten blocks. By reversing the question, the teacher gives struggling students a greater chance at success. In this case, the students are free to use any 16 blocks they wish; all they have to do is figure out the name for the number they have represented.

In the discussion resulting from the task, students will have opportunities to consider important place value ideas, for example, how 16 could be represented by 16 blocks (16 ones) or only 7 blocks (1 ten and 6 ones), but how 448 cannot be represented by fewer than 16 blocks. Some students may begin to realize that one can sum the digits of a whole number to determine the minimum number of blocks needed to represent it. However, even the student who does not recognize this can successfully respond to the task and then will have the opportunity to learn more advanced ideas about place value while listening to other students.

Variations. An obvious variation of this question is a change in the number of blocks mentioned in the question. Other changes might involve asking students to select a number and look for a pattern in the list of the numbers of different base ten blocks that can be used to represent it. They might notice that the values of those numbers always differ by nines: for example, 121 can be represented by 4, 13, 22, 31, . . . blocks.

✳ **BIG IDEA.** **It is important to recognize when each operation (addition, subtraction, multiplication, or division) is appropriate to use.**

> About how many days have you been in school? Tell how you estimated. Tell what mathematical operations you used.

This open question is likely to pique the curiosity of most students. It forces them to consider how they will collect the data and also to consider what computations they will use to answer the question.

Some students will simply focus on the current year. They might think that because there are about 4 weeks in a month and 5 days in most school weeks, they could estimate 20 school days a month. They would then either add or multiply to determine the number of months that have elapsed in the school year.

Others will consider all the years they have been in school. Still others will strive to be more exact, reducing the totals to account for school holidays. Whatever their approach, it is likely that all students will be engaged in the task, and all will be involved in mathematical thinking.

In the discussion of what operations the students used, they should be encouraged to justify why the particular operations were appropriate.

Open Questions for Grades 3–5

> Create a sentence that uses each of these numbers and words:
>
> *21, 84, share, almost*

The sentence approach required in this question allows for a great deal of creativity. For example, students might write:

- *If 4 people share almost 84 cookies, they each get almost 21 cookies.*
- *I wanted to share my 21 jellybeans with my almost 84 friends and then realized that would be a problem; I couldn't split the jellybeans.*
- *If 21 people share almost $84, they each get almost $4.*

A student still struggling with multiplication and division who does not notice the relationship between 21 and 84 might use a sentence such as this one: *21 people each took a sandwich from the tray that the almost 84 people at the party had to share.* Although many students will use division, not all will.

Variations. The words and numbers that are required can easily be changed to allow for other possibilities. Examples of other words or phrases are: *product, quotient, sum, about, times as many.*

✳ **BIG IDEA.** **There are many different ways to add, subtract, multiply, or divide numbers.**

> Make up a subtraction question where the digits 4, 6, and 9 appear somewhere in the question.

All students who respond to the question will use subtraction, but students can choose particular combinations with which they are more comfortable. For example, one student might write: 49 − 6 = 43; another might use **regrouping** and respond: 46 − 9 = 37; and yet another student might write: *9 − 6 − 4 is 1 below 0.*

Discussion of an open question such as this one provides an opportunity for students to engage in computational practice as they check each other's solutions.

Variations. The question can easily be varied by using different sets of three numbers or by using a different operation (e.g., multiplication).

TEACHING TIP. Teachers need to provide a significant amount of computational practice so students will gain procedural fluency. Rather than assigning large numbers of similar questions, posing one open question can efficiently create opportunities to practice in the context of lively class discussion.

Replace the boxes with values from 1 to 6 to make each problem true. You can use each number as often as you want. You cannot use 7, 8, 9, or 0.

This question requires students to multiply two-digit numbers by single-digit numbers but gives them a choice about what numbers to use.

Students who have more difficulty with multiplying may choose to multiply only by 1 and 2 to make the multiplication easier (e.g., $34 \times 1 = 34$, $42 \times 1 = 42$, $62 \times 2 = 124$). Other students might try other combinations, having to take into account that the products must involve only certain digits. For example, a student might begin by multiplying 4 by 2 and realize that this is not allowed because the product is 8, a digit that cannot be used in this case.

All students will have the chance to practice multiplication, using whatever strategies they prefer.

TEACHING TIP. If a mathematically strong student always seems to select values that make a question too easy, the teacher should allow the student to make the initial choice, but then challenge him or her to try other values as well.

OPEN QUESTIONS FOR GRADES 6–8

✷ BIG IDEA. There are many ways to represent numbers.

Describe 100 thousand in as many ways as you can.

Typically this question would be presented using the numeral 100,000, rather than the written phrase *100 thousand*. The advantage of the suggested presentation is that the question becomes more accessible to students who struggle with numerals that represent large numbers. Although one might think that students need to have a concept of how much 100,000 is to answer the question, which would certainly be preferable, a student could simply say that it is more than 99 thousand.

Other students might refer to 100,000 by using place value ideas, for example, indicating that it is possible to represent 100,000 with 100 large base ten blocks, that it is $\frac{1}{10}$ of 1 million, or that it can be represented with 1,000 hundred blocks.

The class discussion would provide ample opportunity for exploration as different aspects of 100 thousand are raised.

> Fill in values for the blanks to make this statement true:
>
> 72 is ____% of ____.

Typically, a teacher might ask a student what a given percentage of a given number is, often with a focus on a particular type of percentage, such as a whole number percentage between 0% and 100%, a fractional percentage, or a percentage greater than 100. This more open question allows the student to choose both the percentage and the number of which the percentage is taken, making the calculation as simple as needed.

Many students who would have difficulty with a question such as *What is 120% of 60?* or *What is the number that 72 is 120% of?* might find it much easier to respond with any of the following:

- *72 is 100% of 72.*
- *72 is 72% of 100.*
- *72 is 50% of 144.*

Even these students can experience success in answering the question. It is likely that in a class of 30 students, many answers will be offered, allowing students the opportunity to consider a wide range of situations.

Variations. The number 72 can be changed. Another possibility is to pose a question such as one of these instead: ____ *is* ____% *of 50* or ____ *is 40% of* ____.

> Create a sentence that uses each of the following words and numbers. Other words and numbers can also be used.
>
> *40, percent, most, 80*

This open question allows for diverse responses. For example, students might write:

- *Most people know that 40 is 50 percent of 80.*
- *80 percent of 40 is most of the 40.*
- *90 percent of 40 and 90 percent of 80 are most of 40 and 80.*

The numbers 40 and 80 were chosen to make the question more accessible. Most students are comfortable with the concept of 50%.

Variations. The words and numbers that are required can easily be changed to allow for other possibilities. Examples of other words are: *ratio, product, added, almost.*

✳ **BIG IDEA.** **Numbers tell how many or how much.**

> The **square root** of a number is easy to figure out. What might the number be?

This question is only appropriate once students have some concept of what a square root is. Whether the calculation of a particular square root is easy or not is, of course, always in the eye of the beholder. Phrasing the question in this way allows the student to take ownership of the problem; it is the individual student who can decide whether a calculation is easy or not.

One student might think that calculation of $\sqrt{144}$ might be easy, just because he or she happens to know that $12 \times 12 = 144$. Many students will suggest that calculating $\sqrt{400}$ is easy because they recognize that if you square a multiple of 10, the result has two zeroes at the end. Most students would suggest that $\sqrt{1}$ or $\sqrt{4}$ is easy to calculate.

In the class discussion, it is likely that a large number of calculations of square roots will be discussed, benefiting all students in the class.

> A number can be written 0.24242424.... What do you know about the size of the number?

Repeating decimals cause difficulty for many students, yet toward the end of this grade band, it is important for students to understand the amounts these numbers represent.

The question is made more accessible to students by phrasing it in the more general sense of *What do you know about the size of the number?* rather than the more specific *What fraction does this number represent?* Even if a student does not know that this decimal is equivalent to $\frac{24}{99}$ or how to figure out the fraction, the student could still indicate that the number is greater than 0.2, that it is less than 25%, or that it is less than 1. In this way, every student in the class can provide a meaningful answer to the question.

It is likely, in the course of discussing the answers, that all students will be exposed to the fraction equivalent for the decimal and a method to show how they could determine that equivalent.

Open Questions for Grades 6–8

✳ **BIG IDEA.** **Number benchmarks are useful for relating numbers and estimating amounts.**

> Choose a fraction and a percent. Tell which is greater and how you know.

It is important that students have an opportunity to consider different representations of **rational numbers**. Two important representations are fractions and percentages.

Rather than directing which rational number and which percentage to compare, this more open question allows the student to choose values. In this way, some equivalence of representations must be considered.

A student who is less comfortable with fraction–percentage relationships might choose the fraction $\frac{1}{1,000}$ and the percentage 99%. It is possible to compare these values without actually changing one representation to the other because one is clearly very close to 0 and the other very close to 100%, or 1. Another student might use fractions such as $\frac{1}{3}$ and percentages such as 40%. Here a student would probably need to know that $\frac{1}{3}$ is about 33%.

As various responses are discussed, the student with less knowledge about converting fractions to percentages or vice versa has the opportunity to gain that knowledge, but also has the opportunity to correctly respond to the question right away.

> Choose something you have been wanting to buy that costs more than $50. Imagine you have $30 saved. What discount does the store need to offer before you can afford it?

This problem will be motivating to many students because it is personal. Students choose the item to buy, so it can be something of particular interest to them.

By being allowed to select a price that is more than $50 rather than exactly $50, students are provided even more choice. This will allow them to select a value that will make the calculation manageable. For example, a student might choose a price of $60 so that a 50% discount would immediately get to the $30 available.

Another student might select a price of $300. That student would then have to recognize that the $30 saved is only 10% of the price and realize that the discount needs to be 90%. Yet another student might use a value of $120 and apply a 75% discount.

Variations. The values $50 and $30 can be changed to create subsequent problems or to make the problem either simpler or more complex.

✷ **BIG IDEA.** By classifying numbers, conclusions can be drawn about them.

> How are the numbers 6.001 and 1.006 alike? How are they different?

Responses to this open question might involve how the numbers can be represented on a place value chart or with base ten blocks, their relative sizes, or simply their digits.

Whereas some students will concentrate on the fact that both numbers use the same digits (thus classifying the numbers in terms of the digits that make them up), others will recognize that one is greater than 6 and one less than 6 (thus classifying the numbers in terms of their relationship to 6), that both involve decimal thousandths (thus classifying the numbers in terms of their decimal representations), or that both are less than 10. Even a struggling student will recognize that the same digits are involved.

Variations. In the example cited, decimal numbers were compared and contrasted. New questions can be created in which pairs of fractions, mixed numbers, or large numbers such as 1.1 million and 1.1 billion are compared.

> 4 is a **factor** of two different numbers. What else might be true about both of the numbers?

A critical aspect of mathematical thinking is the recognition that knowing one thing about a number can provide other information about that number. For example, knowing that a number is greater than 100 automatically means it is also greater than 10.

By knowing that a number is a **multiple** of 4 (because 4 is its factor), a student should be able to tell that the number is even and that it ends in 0, 2, 4, 6, or 8. Some students might even realize that if one subtracts 4 from the number, or adds 4 to the number, the result is also a multiple of 4. It is up to the teacher, by prompting other students, if necessary, to elicit the idea that this is still a **conjecture**. There may be students who can explain why the sum or difference must be a multiple of 4.

With the question posed in the way indicated, a student who is less likely to generalize can still select two specific numbers and tell something true about both of them. For example, the student might choose 8 and 12 and point out that both are even.

Variations. Similar questions can be posed by changing the number that is the factor.

✳ **BIG IDEA.** The patterns in the place value system can make it easier to interpret and operate with numbers.

> Suppose that the numbers x and y are both decimals. You know that $x = 1,000y$. What might x and y be?

One of the important ideas about the place value system is that the places are related by **powers of 10**. If there is a 4 in a particular spot in a whole or decimal number, that same 4 one space to the left is worth 10 times as much, two spaces to the left is worth 100 times as much, and so on.

With this open question, students are being asked to recognize that x and y must have the same digits, just shifted over. By not specifying a value for either x or y, the teacher could have one student who chooses 3.0 and 3,000.0, and another who chooses 0.003 and 0.000003.

Discussion of the wide range of responses will enhance and reinforce the understanding of pattern in the place value system for students at all levels.

✳ **BIG IDEA.** It is important to recognize when each operation (addition, subtraction, multiplication, or division) is appropriate to use.

> You divide two numbers and the answer is 2.5. What two numbers might you have divided? What word problem might you have been solving?

With this open question, students can choose to use either whole numbers or decimals for the **dividend** and the **divisor**, and they can choose what sort of context to use for their word problems. This choice will make the question accessible to a broad range of students.

Some students might choose $2.5 \div 1$, or, if they recognize that $2.5 = 25$ tenths, they might choose values 25 and 10, because $25 \div 10 = 2.5$. Other students will use more varied calculations, for example, $5 \div 2$, $50 \div 20$, $0.5 \div 0.2$, $100 \div 40$, and so on. Word problems can cover a wide range of topics and degrees of complexity, such as:

- *I divided a string that was 12.5 m long into 5 pieces. How long was each piece?*
- *I had 25 pounds of hamburger meat. I was making packages of 10 pounds. How many packages could I fill?*

The variety of answers students offer will make it easy to talk about the fact that the quotient does not change if both the divisor and dividend are multiplied by the same number. For example, $10 \div 4$ produces the same result as 20 (i.e., 2×10) $\div$ 8 (2×4) or 5 ($10 \div 2$) $\div$ 2 ($4 \div 2$).

> Choose a price for 4 cinnamon buns. Then choose a different number of cinnamon buns and tell how much that new number of buns would cost. Tell how you know you are correct.

By allowing students to choose both the price and the number of cinnamon buns to be used, this problem is suitable for a broad range of students.

Some students will select a simple price (e.g., $4), whereas others will select a more complicated price (e.g., $2.79). Some students will select an easy number of buns (such as 8 or 12, multiples of 4), whereas others will select a more complicated number (such as 3, not a multiple of 4).

By allowing—and discussing—many different solutions, each student in the class will be exposed to how to solve many rate problems.

✳ **BIG IDEA.** **There are many different ways to add, subtract, multiply, or divide numbers.**

> Create a question involving multiplication or division of decimals where the digits 4, 9, and 2 appear somewhere.

All students who respond to the question will use multiplication or division, but different students can choose particular combinations with which they are more comfortable. For example, a student might use $94.0 \div 2.0$ or $2.0 \times 4 \times 9$ to employ whole numbers. Other students more comfortable with decimals might create questions such as $9.2 \div 4$, 9.2×0.4, $36.8 \div 4$ (where the quotient is 9.2), and so on.

Variations. To make the questions even more open, the teacher might remove the requirement for decimal multiplication and division, but instead circulate and personally encourage some students to use decimals.

TEACHING TIP. One way to open up a question is to remove phrases that require a certain type of number to be used in the calculation.

> You divide two fractions, and the numerator of the quotient is a 4. What could the fractions be?

Students can consider one or more combinations of fractions that might be divided and can use any algorithm or procedure with which they are comfortable.

Students who divide fractions by dividing the numerators and dividing the denominators (e.g., $\frac{8}{15} \div \frac{2}{5} = \frac{8 \div 2}{15 \div 5}$) will find this question easier than other students. But almost any student can experience some success simply by using fractions such as $\frac{4}{5}$ and $\frac{1}{1}$ or $\frac{4}{9}$ and $\frac{1}{1}$. Or a student might propose dividing $\frac{8}{5}$ by $\frac{2}{1}$ or $\frac{8}{10}$ by $\frac{2}{1}$.

By multiplying each of the divisors by 4, the teacher can reinforce the relationship between multiplication and division, whether of fractions or whole numbers.

Variations. The quotient can be changed, as can the operation. For example, the question might involve multiplying or adding fractions.

✸ **BIG IDEA.** It is important to use and take advantage of the relationships between the operations in computational situations.

> You combine two integers to get −2. What integers might you have combined and how?

By using the word *combine* instead of specifying an operation, this problem allows for more student success.

For example, a student who knows how to multiply by −1 might suggest that −1 is combined with 2 by multiplying. Another student might suggest combining −2 and 0 by adding. Yet another student, who knows that that subtracting a positive integer from 0 results in its opposite, might suggest that 0 is combined with 2 by subtracting. Students who are comfortable adding integers might suggest one or more combinations of integers that sum to −2, for example, −3 + 1, −4 + 2, and so on.

Variations. The value of −2 can be altered. In addition, if adding or multiplying is required, the number of integers being combined can vary.

PARALLEL TASKS FOR PREKINDERGARTEN–GRADE 2

PARALLEL TASKS are sets of two or more related tasks that explore the same big idea but are designed to suit the needs of students at different developmental levels. The tasks are similar enough in context that all students can participate fully in a single follow-up discussion.

✸ **BIG IDEA.** There are many ways to represent numbers.

> What coin combinations can you use to show your amount?
>
> *Option 1:* 12¢   *Option 2:* 60¢

Frequently, in a math class, a single task is assigned to all students. For example, all students might be asked to represent 32¢ with coins. By allowing a choice that provides for different levels of complexity, more students can achieve success. The student who is really only comfortable with dimes and pennies can be as successful in his or her task as another student who can deal with a broader range of coins and more combinations.

There are only four combinations that will yield 12¢ (12 pennies, 7 pennies and 1 nickel, 2 pennies and 1 dime, or 2 pennies and 2 nickels), whereas many more combinations will produce 60¢ (e.g., 6 dimes; 5 dimes and 10 pennies; 5 dimes, 1 nickel, and 5 pennies; 2 quarters and 10 pennies; etc.).

Questions that would be appropriate to ask all students, no matter which task they chose, include:

- *Did you use any dimes?*
- *Did you need to use pennies?*
- *What coins did you use?*
- *Were there other possible combinations?*

TEACHING TIP. When numbered options are offered, sometimes the "simpler" option should be presented as Option 1 and other times as Option 2. The unpredictability will ensure that students consider both possibilities when they choose their tasks.

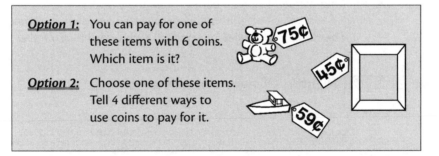

Option 1: You can pay for one of these items with 6 coins. Which item is it?

Option 2: Choose one of these items. Tell 4 different ways to use coins to pay for it.

Students who pursue **Option 1** are able to solve a problem involving representing numbers many different ways. They need to recognize, for example, that even though 45¢ can be represented with 9 coins (4 dimes and 5 pennies), it can also be represented with 6 coins (3 dimes and 3 nickels). They must consider all the options for representing each amount.

Those who pursue **Option 2** need only to be able to represent a single number in different ways; with the freedom to choose the number to represent, an even greater likelihood of success for each student is ensured. A student might, for example, choose the 75¢ teddy bear and suggest it could be paid for by using 75 coins (all pennies), 3 coins (3 quarters), 12 coins (7 dimes and 5 pennies), or 5 coins (2 quarters, 2 dimes, and 1 nickel).

Questions to ask all students, no matter which task they chose, would include:

- *What item did you choose?*
- *Why did you choose that item?*
- *Could you have used two quarters?*
- *How many coins did you use?*
- *How did you figure out what coins to use?*

> **Option 1:** Choose a number between 1 and 10. Show that number in as many ways as you can.
>
> **Option 2:** Choose a number between 20 and 30. Show that number in as many ways as you can.

The difference between the two options is very slight, but providing a choice allows students who are working at different mathematical levels to select the more appropriate task.

Students might represent their numbers by using pictures, numbers, or words. For example, a student might represent 9 as $10 - 1$, as a picture of 9 items, as 9 dots in a 10-frame, as $4 + 5$, as 3 groups of 3, or by using stylized versions of the numeral 9.

No matter which task was chosen, students could be asked:

- *What number did you represent?*
- *How do you know that that number is one that was okay to choose?*
- *What are some of the different ways you represented that number?*

✸ **BIG IDEA.** Numbers tell how many or how much.

> **Option 1:** Choose a number that could tell how many flowers might fit in a vase. Tell why that number makes sense.
>
> **Option 2:** Choose a number that could tell the number of families in a small town. Tell why that number makes sense.

The number of flowers that fits in a vase is likely to be a fairly small number. However, there is some flexibility. A student might think of a larger vase that might hold 20 or more flowers. The number of flowers is unlikely to be in the hundreds, however.

The number of families in a small town is likely to be a two-digit, three-digit, or maybe even a four-digit number, depending on the individual student's personal view of what constitutes a small town. The discussion would be an interesting one as students share their personal views on what numbers are reasonable and why.

All students could be asked how they decided on their numbers and why they thought their numbers were reasonable.

✳ **BIG IDEA.** **Number benchmarks are useful for relating numbers and estimating amounts.**

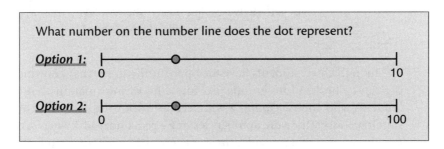

These parallel tasks allow students who are working with only small numbers to succeed at a task similar to the one used by students able to work with greater numbers. The same strategies are likely to be used no matter which number line is chosen. Most students will mentally divide the line in half, then that half in half, and so on.

Variations. The task can be varied by using alternative endpoints for the number line and alternative positions for the dot.

> **Option 1:** A number is about 10, but it's not 10. What is the most it might be? What is the least it might be?
>
> **Option 2:** A number is about 125, but it's not 125. What is the most it might be? What is the least it might be?

Again, the difference between the two options is very slight, but it allows students who are working at different mathematical levels to choose options more appropriate for them.

It is important for students to recognize that the word *about* in the question is vague yet can still be meaningful. They will probably find it interesting that the larger the number, the wider the range of values that will be regarded as appropriate estimates. For example, for numbers about 10, the range might be about 8 to 12; for 125, however, the range might be about 115 to 135.

✷ **BIG IDEA.** **By classifying numbers, conclusions can be drawn about them.**

> **_Option 1:_** Ian had some markers. When he put them in groups of 3, there were 2 left over. If he had fewer than 15 markers, how many could he have had?
>
> **_Option 2:_** Andrea had some markers. When she put them in groups of 3, there was 1 left over. When she put them in groups of 4, there were 3 left over. If she had fewer than 20 markers, how many could she have had?

In both cases, students have the opportunity to see that knowing how a number can be broken into groups provides a lot of information about the number. Students who choose **_Option 2_** will need to work with two pieces of information simultaneously (the only numbers leaving a remainder of 1 when divided by 3 and remainder of 3 when divided by 4 are 7 and 19), whereas students who choose **_Option 1_** have a simpler but equally worthwhile task (5, 8, 11, and 14 are options).

Questions such as the following would be valuable no matter which task students completed:

- *Could your number have been 5? 15?*
- *How did you decide what numbers you could try?*
- *How did you begin to solve the problem?*

✷ **BIG IDEA.** **It is important to recognize when each operation (addition, subtraction, multiplication, or division) is appropriate to use.**

> Make up a word problem for one of these equations:
>
> **_Option 1:_** $38 + 26 = 64$ **_Option 2:_** $3 \times 8 = 24$

Students can create a problem for either addition or multiplication. Many students will find the addition problem easier even though the numbers are greater, but students can choose whichever is more comfortable for them. For either option, students could be asked how they decided on what their problem would be about and how they knew that their problem matched the number sentence they had chosen.

✳ **BIG IDEA.** There are many different ways to add, subtract, multiply, or divide numbers.

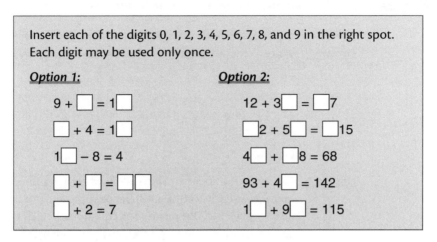

Insert each of the digits 0, 1, 2, 3, 4, 5, 6, 7, 8, and 9 in the right spot. Each digit may be used only once.

Option 1:

$9 + \square = 1\square$

$\square + 4 = 1\square$

$1\square - 8 = 4$

$\square + \square = \square\square$

$\square + 2 = 7$

Option 2:

$12 + 3\square = \square 7$

$\square 2 + 5\square = \square 15$

$4\square + \square 8 = 68$

$93 + 4\square = 142$

$1\square + 9\square = 115$

Missing digit tasks are excellent tools for providing practice for students in both computation and reasoning. In either case shown above, students practice addition and subtraction skills. Students who choose _Option 2_ are comfortable working with two-digit numbers, whereas students more comfortable with simple addition facts can choose _Option 1_. _Option 1_ may actually be viewed as more complex because there are many possible combinations for each of the number sentences.

For example, notice that only the third and fifth expressions for _Option 1_ have one possible solution, whereas for _Option 2_ there is only one possible solution for each of the first four number sentences. A solution for _Option 1_ uses the following numbers from top to bottom, left to right: 8, 7, 9, 3, 2, 4, 6, 1, 0, 5; a solution for _Option 2_ uses these: 5, 4, 6, 3, 1, 0, 2, 9, 7, 8.

Whichever task students performed, they could be asked which sentence they worked on first and why, which values were easiest for them to find and which more difficult, and what their solutions were.

TEACHING TIP. Students should occasionally be encouraged to explain why they chose the options they did.

Parallel Tasks for Pre-K–Grade 2

PARALLEL TASKS FOR GRADES 3–5

✷ **BIG IDEA.** There are many ways to represent numbers.

> ___Option 1:___ Write as many equations as you can to describe this **array**.
>
> ```
> x   x   x   x   x   x   x   x   x
> x   x   x   x   x   x   x   x   x
> x   x   x   x   x   x   x   x   x
> x   x   x   x   x   x   x   x   x
> x   x   x   x   x   x   x   x   x
> ```
>
> ___Option 2:___ Can every number from 1 to 50 be written in more than
> one way as each: a sum, a difference, a product, or a
> quotient? Write your answers as equations. Which was
> easiest: sums, differences, products, or quotients?

Some students will need the visual stimulus of an array to help them see how the number 45 can be written in terms of other numbers. Looking at the array, most students will see sums and products; they can be challenged to look for differences and quotients as well. Rather than specifying multiplication equations (such as $5 \times 9 = 45$), the question is phrased more broadly, asking simply for equations. This allows an entry point even for students still not comfortable with multiplication (e.g., $9 + 9 + 9 + 9 + 9 = 45$).

Other students will be prepared to work more symbolically. They might select ___Option 2___, which encourages them to make conjectures and then test them. They may come to realize that there is an easy way to write every number as a sum ($x + 0$), a difference ($x - 0$), a product ($1 \cdot x$), or a quotient ($x \div 1$). And they may discover that coming up with alternative ways to obtain some numbers by writing a product will not be as easy as coming up with alternatives for the other operations.

In follow-up discussion, all students could be asked what their equations were and how they arrived at them.

> ___Option 1:___ Two fractions are equivalent. If you add the numerators,
> the result is 22 less than if you add the denominators.
> What could the fractions be?
>
> ___Option 2:___ Draw a picture to show two equivalent fractions for $\frac{2}{8}$.

The fact that a part of a whole can be represented in many ways is fundamental to students' ability to add and subtract fractions. Although many students learn the rule for creating equivalent fractions, some use it without understanding why it works; others generalize the rule inappropriately, for example, adding the same amount to both numerator and denominator instead of only multiplying or dividing by the same amount.

The choice of tasks shown for this exercise allows the student who is just beginning to understand equivalence to show what he or she knows about why two fractions might be equivalent. It also allows the more advanced student to work in a more symbolic way to solve a problem involving equivalence. Because there are many solutions to *Option 1* (e.g., $\frac{2}{4} = \frac{20}{40}$ and $(40 + 4) - (20 + 2) = 22$; $\frac{6}{10} = \frac{27}{45}$ and $(10 + 45) - (6 + 27) = 22$; $\frac{8}{10} = \frac{80}{100}$ and $(10 + 100) - (8 + 80) = 22$), students must recognize that they should try more than one combination of values.

Questions that could be asked of both groups include:

* *What two equivalent fractions did you use?*
* *How did you know they are equivalent?*
* *What kind of picture could you draw to show that they are equivalent?*
* *How did you solve your problem?*

✴ BIG IDEA. **Numbers tell how many or how much.**

> What real-life situations might this number describe?
>
> ***Option 1:*** 10,000   ***Option 2:*** 1,000

Some students in grades 3–5 will find the number 10,000 difficult to relate to. Even if they can read it, they often have no sense of its size. Other students who think more proportionally can imagine 10,000 in terms of known numbers, for example, 10 thousand cubes or the number of people a local arena holds. By providing options for which number to work with, all students are more likely to benefit from the task.

Variations. A different pair of numbers, whether two other whole numbers, two decimals, or two fractions, can be used instead of the pair above.

TEACHING TIP. One of the benefits of parallel tasks is that even though a student may choose an "easier" option, he or she still benefits from class discussion of the other option.

Parallel Tasks for Grades 3–5

> **_Option 1:_** Twice as many people came in ahead of David's dad as behind him in a marathon run. If there were 112 participants, what was David's dad's position?
>
> **_Option 2:_** Twice as many people came in ahead of David's dad as behind him in a marathon run. What are the possible numbers of runners?

Both of the suggested tasks are suitable problems for students in grades 3–5. The tasks differ in that the first task is very specific and is solvable with many strategies, even guess and test. The second task, which requires a generalization, forces a student to think more analytically. He or she is likely to try a variety of numbers and then look for commonalities before noticing that the number must be one more than a multiple of 3.

Questions that could be asked of both groups of students include:

- *Could there have been 50 runners ahead of David's dad? How do you know?*
- *How do you know that if David's dad came in 25th, there would not be 15 people behind him?*

> Choose one of the measurements shown below. About how many years old is someone who is as old as the measurement you chose?
>
> - 1,000 days   • 10,000 hours   • 1,000,000 seconds

Students presented with this problem have the choice of whether to work with a very large number (1,000,000) or with smaller numbers (1,000 or 10,000) that might be more comfortable for them. No matter which choice students make, they practice computations involving time measurement conversions (1,000 days is slightly less than 3 years; 10,000 hours is about 14 months; 1,000,000 seconds is between 11 and 12 days); the choice simply allows them to deal with conversions that are more accessible to them. No matter which choice students make, there is still a puzzle to be solved, one that is likely to be of interest.

TEACHING TIP. One of the most important things a teacher can do for struggling learners is to ensure that they have the opportunity to enjoy the puzzle aspect of math, but at a level at which they can succeed. Providing only rote tasks will never give students the opportunity to truly enjoy math, even if they do enjoy the success of being correct.

Parallel Tasks for Grades 3–5

✳ **BIG IDEA.** **Number benchmarks are useful for relating numbers and estimating amounts.**

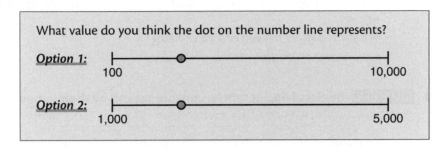

What value do you think the dot on the number line represents?

Option 1:

100   10,000

Option 2:

1,000   5,000

These parallel tasks allow some students to work with a smaller number range and smaller numbers if they wish. Students who choose *Option 2* are likely to mentally divide the line into fourths and see the dot as being at about 2,000. Those who pursue *Option 1* are likely to estimate the 100 as 0 to help them mentally divide the line.

In discussing the work of all students, a teacher might ask:

- *Is your dot worth more than 1,000? Why or why not?*
- *Is it worth more than 2,000? Why or why not?*
- *How did you figure out the value the dot represents?*
- *How do you know that there has to be more than one reasonable answer to the question?*

Variations. The task can be varied by using alternative endpoints for the number line and alternative positions for the dot.

Option 1: You multiply two numbers and the product is about 40 more than 42 × 63. What numbers might you have multiplied?

Option 2: You multiply two numbers and the product is close to 2,600. Both numbers are greater than 10. What could they be?

Both of these tasks involve students in thinking about multidigit multiplication. A student who completes *Option 1* might realize that 42 × 64 is 42 more than 42 × 63 and thus would satisfy the conditions of the problem. Alternatively, a student might calculate the product as 2,646, add 40 to get 2,686 and select numbers such as 3 × 900.

A student who completes *Option 2* might begin with 26 × 100, and then select a response of 25 × 100 to give a product is close to 2,600 but not exactly that number.

A teacher could ask all students questions such as:

- *Could both of your numbers be greater than 60? Greater than 50?*
- *Suppose one of your factors was about 20. What would the size of the other one be?*
- *What two factors did you end up choosing?*
- *How do you know they work?*

✸ **BIG IDEA.** **By classifying numbers, conclusions can be drawn about them.**

> ***Option 1:*** The sum of the digits of a three-digit number is 12. How many base ten blocks could you use to model that number?
>
> ***Option 2:*** You can model a number with 58 base ten blocks. What other numbers of base ten blocks could you use to model that number?

One of the important principles involved in modeling a number with base ten blocks is that for numbers greater than 10, there is always more than one way to represent the number with blocks. For example, 32 can be represented by 5 blocks (3 tens + 2 ones), 14 blocks (2 tens + 12 ones), 23 blocks (1 ten + 22 ones), or 32 blocks (32 ones). Another important idea is that the minimum number of blocks required to model a number is the sum of the digits of the number (e.g., 3 + 2 = 5 blocks for 32). Both of the task options above get at these ideas, but at different levels of sophistication.

The alternative numbers of blocks that can be used to represent a number differ by multiples of 9. This is because when trading blocks, 1 large block is lost for each 10 small ones with which it is replaced; the end result is an increase of 9 blocks. Thus, a number that can be represented with 58 base ten blocks might also be represented, for example, with 22 blocks (because 58 − 22 is a multiple of 9). And this 22-block number would have a digit sum of 22 (e.g., 688, which can be represented by 6 hundreds + 8 tens + 8 ones = 22 blocks, or by 2 hundreds + 48 tens + 8 ones = 58 blocks).

Common questions could be asked of both groups of students, for example:

- *What number did you use?*
- *How did you know that the number was correct?*
- *Could you have used a different number? How do you know?*
- *How many other blocks could you have used to represent your number?*

Variations. For either option, numbers can be changed to make the tasks more or less challenging. For ***Option 1***, students can be given a choice of working with two-digit or three-digit numbers, and they can be asked to provide more than one solution. For ***Option 2***, students can be asked to state a generalization that describes their findings.

✳ **BIG IDEA.** It is important to recognize when each operation (addition, subtraction, multiplication, or division) is appropriate to use.

> _Option 1:_ There are 583 students in Amy's school in the morning. 99 of the third-grade students leave for a field trip. How many students are left in the school?
>
> _Option 2:_ There are 61 third-grade students in Amy's school. 19 of them are in the library. How many are left in their classrooms?

Two similar problems are offered, but one involves three-digit numbers and the other two-digit numbers. By completing either task successfully, students can show their understanding of when and how to subtract. **_Option 1_** is set up to allow for very simple mental math (583 − 99 is 1 more than 583 − 100), so even students who are not confident in three-digit subtraction can still be successful.

All students could be asked:

- _How did you calculate the number of students who were left?_
- _What operation did you use?_
- _Why did you use that operation?_

> _Option 1:_ Create a word problem that can be solved by subtracting two large numbers.
>
> _Option 2:_ Create a word problem that can be solved by subtracting two fractions.

Both options require students to recognize situations where subtraction is appropriate, but one requires the use of fractions and the other allows for whole numbers. For many students, contexts involving fractions are more problematic.

A teacher could ask all students in the class questions such as:

- _Why does your problem involve subtraction?_
- _Could it be solved using another operation? Which one?_
- _What number sentence would go with your problem?_
- _What is your problem?_
- _What is the solution to your problem?_

Parallel Tasks for Grades 3–5

✳ **BIG IDEA.** There are many different ways to add, subtract, multiply, or divide numbers.

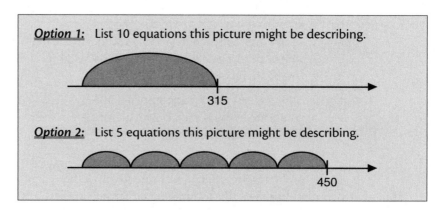

Although some students might assume that **Option 2** requires multiplication, it can involve either addition or multiplication or both (e.g., $5 \times 90 = 450$ or $250 + 5 \times 40 = 450$). **Option 1** is likely to be viewed as an addition problem, although subtraction can also be used to determine solutions (e.g., $100 + 215 = 315$ or $315 - 100 = 215$).

These tasks are suitable as parallel tasks because both involve the use of the "empty" number line and both involve creating simple equations. The difference between the tasks involves the complexity of the operations required.

The ensuing discussion would be a rich one. Students might see that for **Option 2**, they could choose any value repeatedly for the jump as long as they start at an appropriate value. For example, each jump could be 90 if the jumps start at 0, but each jump could be 80 if they begin at 50. Students would find that they could use any two numbers that add to 315 for **Option 1**, so there would be many possible correct number sentences. In each situation, a teacher could ask:

- *What was your starting point?*
- *How did you figure it out?*
- *What operations did you use?*

Insert each of the digits 0, 1, 2, 3, 4, 5, 6, 7, 8, and 9 in the right spot. Each digit may be used only once.

Option 1:

$3 \times 5\square = 15\square$

$\square \times \square\square = \square 0$

$6 \times 1\square = 6\square$

$\square \times 1\square = 105$

Option 2:

$\square \times \square = 5\square$

$4 \times \square = 3\square$

$\square \times \square = 2\square$

$8 \times \square = 4\square$

Students can choose to work with single-digit products or products involving two-digit numbers, whichever is more suitable for them. For either task, it would be instructive to ask which digits they figured out first and why those digits were the easiest ones to get. Each task is a challenge, just a different sort of challenge. When offered a choice, many more students will have the opportunity for success. (*Solutions: **Option 1**,* top to bottom, left to right: 3, 9, 4, 2, 0, 8, 1, 6, 7, 5; ***Option 2**:* 9, 6, 4, 8, 2, 7, 3, 1, 5, 0.)

PARALLEL TASKS FOR GRADES 6–8

✷ **BIG IDEA.** There are many ways to represent numbers.

> **Option 1:** Show −12 as the sum of other integers. What is the largest value you can use for the other integers? What is the smallest value?
>
> **Option 2:** Show −12 as the product of other integers. What is the largest value you can use for the other integers? What is the smallest value?

Even though students generally learn about products of integers after learning about sums, many students might find **Option 2** easier than **Option 1**. In each instance, students must think about the sign rules that apply to integer operations. They also use reasoning in deciding the largest and smallest values that can be used. For example, a student might realize that −12 could be the sum of 400 and −412 or 1,000 and −1,012, recognizing there is no smallest or largest possible value to create a sum. On the other hand, if only integers are used, the smallest value that can be used to form the product is −12 and the largest is +12.

Questions that could be asked of both groups include:

- *How did you write −12?*
- *How many combinations did you use?*
- *What was the largest value you used?*
- *How do you know whether or not a larger value is possible?*

> Choose one of the numbers below. Show at least four or five ways to represent it.
>
> -8 1,000,000,000 0.002

The students have three options for this task. Some will find representing -8 easiest because 8 is a relatively small number. However, these students will have to be comfortable with negative integers. A simple position on a number line would provide one representation of the number. Other methods might involve a model using colored chips or expressions such as $0 - 8$ or $8 - 16$.

Students who choose 1 billion must be comfortable with large numbers. By employing the place value system, students could think of the number as 1,000 millions or 1,000,000 thousands. It could even be regarded as the approximate population of India.

Students who choose 0.002 must be comfortable with decimal thousandths. Dealing with decimal thousandths is considerably more difficult for students than working with decimal tenths or hundredths. Students might show 2 small cubes in a base ten block set relative to a single large cube, might fill in 2 small rectangles on a **thousandths grid**, or might think of 0.002 as 0.2 hundredths.

Students could post their various representations and the teacher could ask:

- *How could you show your number using addition or subtraction? Multiplication or division?*
- *What everyday life object or situation might be described by your number?*

✸ **BIG IDEA.** Numbers tell how many or how much.

> ***Option 1:*** A number between 20 and 30 is 80% of another number. What could the second number be?
>
> ***Option 2:*** A number between 20 and 30 is 150% of another number. What could the second number be?

In both situations, students must determine the number of which a given number is part. Rather than choosing a specific number, a range is used. This provides greater practice and allows students to use convenient benchmark numbers if they wish. ***Option 1*** uses a percentage less than 100%, whereas ***Option 2*** uses a percentage greater than 100%, a more difficult task for most students.

In both instances, students must think about the relationship of the number in the 20s to the other number, that is, whether the other number is greater or less than the given one and why. A possible solution for ***Option 1*** is that 24 is 80% of 30; a possible solution for ***Option 2*** is that 24 is 150% of 16.

Questions that could be asked of both groups include:

- *Is the second number greater or less than the first one?*
- *How did you decide whether the second number was greater or less than the first one?*
- *How far apart are the possible values for the second number?*

Option 1: On a scale diagram, 1 inch represents 4 miles. If two places are $3\frac{1}{2}$" apart on the map, how far apart are the actual places?

Option 2: On a scale diagram, 1 inch represents $7\frac{1}{2}$ miles. If two places are $3\frac{1}{2}$" apart on the map, how far apart are the actual places?

The different scales used in the two options lead to either simpler or more complex calculations. In **Option 1**, the calculation is fairly simple ($4 \times 3\frac{1}{2} = 14$); students need to be able to multiply a mixed number by a whole number. In **Option 2**, students must multiply two mixed numbers ($3\frac{1}{2} \times 7\frac{1}{2} = 26\frac{1}{4}$). In both cases, students are employing the concept of how a scale can be used to represent amounts.

Both groups of students could be asked:

- *How far apart are the towns?*
- *How did you figure it out?*
- *Why did the scale only have to tell you about 1" and not 2" or 3"?*
- *Which was the easiest part of your calculation?*
- *Which part was the most difficult for you?*

✸ **BIG IDEA.** **Number benchmarks are useful for relating numbers and estimating amounts.**

Option 1: Draw a picture that would help someone compare $\sqrt{130}$ and $\sqrt{260}$.

Option 2: Draw a picture that would help someone compare $\sqrt{121}$ and $\sqrt{484}$.

In both situations, students can use benchmarks, such as $\sqrt{100}$, $\sqrt{225}$, or $\sqrt{400}$, to estimate the required square roots. In **Option 2**, the square roots are perfect squares; in **Option 1**, they are not. Some students might think about the Pythagorean theorem and draw a picture of a triangle; others might simply draw line segments of particular lengths, squares of particular areas, or perhaps a graph. If, for example, a student drew a graph showing $y = x^2$, he or she could locate the squares in question and then look for the corresponding square roots.

A teacher could ask all students, regardless of which task they chose:

- *How did you estimate the square roots?*
- *What easy square roots did you use to help you?*
- *How did your picture help make the relationship clear?*

Option 1: You know that the number π is the ratio of the circumference of a circle to its diameter. How would you help someone understand why π is between 3 and 4?

Option 2: You know that $\sqrt{20}$ is the number you multiply by itself to get 20. How could you convince someone it must be between 4 and 5?

In each situation, students must relate a less familiar value to more familiar whole number values. Students could use numerical arguments (probably more so for **Option 2** than for **Option 1**) or pictorial arguments.

For example, a student might show why π is between 3 and 4 with a circle-in-square diagram like the one shown at the right. The diagram shows that if the side of the square was 1, so would be the diameter of the circle. The perimeter of the square would be 4; the circumference of the circle is clearly less, but not a lot less.

A student might show that $\sqrt{20}$ is between 4 and 5 by drawing a right triangle with leg lengths of 2 and 4. The hypotenuse would have length $\sqrt{20}$ and is clearly only slightly longer than the leg with a length of 4.

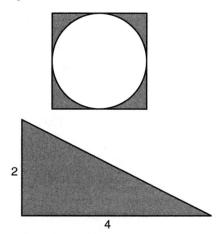

No matter which task students completed, the teacher could ask:

- *Why is it useful to estimate your value using whole numbers?*
- *How does a picture make it easier to estimate the value than just doing calculations?*

✳ **BIG IDEA.** **By classifying numbers, conclusions can be drawn about them.**

> _Option 1:_ Create 10 different patterns that would include the numbers 50 and 1,000,000 among their **terms**.
>
> _Option 2:_ Create 10 different patterns that would include the numbers 1,000 and 10,000 among their terms.

In both situations, students must think about the relationship between a power of 10 (whether 100, 1,000,000, 1,000, or 10,000) and other numbers. This is valuable not only because powers of 10 are important numbers for students to internalize but also because using number patterns such as these helps build number sense. Both tasks are likely to require students to classify numbers, whether as multiples of 50, multiples of 1,000, and so on.

Some students will find _Option 2_ easier because the numbers are simpler and the pattern 1,000, 2,000, 3,000, … is an obvious starting point. Other students would benefit from the challenge of _Option 1_. They could create a pattern such as 50, 100, 150, 200, … as long as they could explain why 1,000,000 would have to be part of the pattern. In this case, students are likely to use more complex number relationships than are required for _Option 2_.

Questions such as these could be asked of either group:

- _What is your beginning number?_
- _What is your **pattern rule**?_
- _How do you know that the pattern satisfies the required conditions?_

> _Option 1:_ Are there more multiples of 3 or more multiples of 4 between 1 and 100? How many more?
>
> _Option 2:_ Are there more **prime numbers** or more **composite numbers** between 1 and 100? How many more?

Classifying numbers in terms of multiplicative properties is an underlying theme in grades 6–8. Whether numbers are being classified as primes and composites or as multiples of other numbers, students are considering multiplicative properties.

Some students at this level still struggle to decide whether a number is prime. These students might be more successful with _Option 1_, yet they would benefit from the ensuing discussion that involves _Option 2_. For either task, students will make two lists of numbers and compare their lengths. In addition, they will need to justify why particular numbers are on one list, the other list, or both lists (in the case of _Option 1_). As the discussion continues, it would be useful to ask why numbers appear on both lists in _Option 1_ but not in _Option 2_. There are $33 - 25 = 8$ more multiples of 3 between 1 and 100 than multiples of 4. There are $74 - 25 = 49$ more composites than primes. The number 1 is neither prime nor composite.

✳ **BIG IDEA.** It is important to recognize when each operation (addition, subtraction, multiplication, or division) is appropriate to use.

> **_Option 1:_** Draw a picture to show why $\frac{2}{3} \times \frac{5}{6} = \frac{5}{9}$.
>
> **_Option 2:_** Draw a picture to show why $\frac{2}{3} \times \frac{7}{3} = \frac{14}{9}$.

One of the struggles students experience in grades 6–8 is comprehending why the rules for multiplying fractions work. The rules themselves are easy to learn, but making sense of them requires conceptual understanding. One way for students to either demonstrate that understanding or acquire it is to draw a model for multiplication, taking advantage of prior knowledge relating the area of rectangles to multiplication.

For **_Option 1_**, students might draw the picture on the left below; for **_Option 2_**, they might draw the picture on the right:

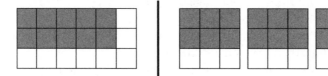

Many students find a diagram illustrating improper fractions much more difficult to create or explain. They often become confused about how many parts actually make up one whole. Therefore, the choice of tasks is provided so that students can work with either proper or improper fractions. In each case, students must explain why their picture demonstrates the algorithm.

✳ **BIG IDEA.** It is important to use and take advantage of the relationships between the operations in computational situations.

> **_Option 1:_** Show that the product of two numbers can sometimes be greater than the quotient and sometimes less.
>
> **_Option 2:_** Choose two numbers to make each statement true:
>
> quotient < difference < sum < product
>
> sum < difference < product < quotient

Students, and even many adults, harbor many misconceptions about multiplying and dividing. They often believe that multiplication always makes a number greater and division makes a number smaller, without considering the complication of using fractions or negative numbers. The task shown above requires consideration of those complicating factors.

Option 1 is less demanding than Option 2 in that fewer conditions are imposed. As a solution to Option 1, a student might offer $-3 \times -4 > -3 \div -4$, but $3 \times 0.5 < 3 \div 0.5$. For Option 2, a student might use the pair of numbers 12 and 2 for the first part and the pair of numbers -12 and -0.5 for the second part.

Questions that could be asked of both groups include:

- *Suppose you were using 25 and 5. Which is greater—the quotient or the product?*
- *Is it ever possible for the quotient of two numbers to be greater than the product? When would that be?*
- *What two numbers did you choose? Why did you try those?*

Option 1: There were 10,625 athletes in the 2004 summer Olympics. Of these, 4,329 were female. Calculate the number and percentage of athletes who were male. Estimate to check your work.

Option 2: 850 athletes participated in the local track and field event for the Special Olympics. Of these athletes, 512 were female. Calculate the number of athletes who were male. Estimate the fraction who were male.

Two similar problems are offered, but one involves numbers that may be difficult for some students to work with. By selecting Option 2, students can still show their understanding of when and how to subtract and to use fractions, but in a situation where they are more likely to be successful. When problems are presented in contexts that are of interest to students, it is even more likely that a broad range of students will be engaged.

All students could be asked:

- *How did you calculate the number of males?*
- *Were more than half males? How do you know?*
- *What operation did you use? Why did you use that operation?*
- *How did you estimate? Why was that estimate appropriate?*

SUMMING UP

The eight big ideas that underpin work in Number and Operations were explored in this chapter through 75 examples of open questions and parallel tasks, as well as variations of them. The instructional examples provided were designed to support differentiated instruction for students at different developmental levels, targeting three separate grade bands: pre-K–grade 2, grades 3–5, and grades 6–8.

Number and Operations remains the most fundamental strand for many teachers and parents; student success in this strand is critical. Applying differentiated instruction strategies in number work will guarantee that many more children will experience success in mathematics.

MY OWN QUESTIONS AND TASKS

Lesson Goal: Grade Level: _____

Standard(s) Addressed:

Underlying Big Idea(s):

Open Question(s):

Parallel Tasks:

Option 1:

Option 2:

Principles to Keep in Mind:

- All open questions must allow for correct responses at a variety of levels.
- Parallel tasks need to be created with variations that allow struggling students to be successful and proficient students to be challenged.
- Questions and tasks should be constructed in such a way that will allow all students to participate together in follow-up discussions.

The examples presented in this chapter only scratch the surface of possible questions and tasks that can be used to differentiate instruction in Number and Operations. Other tasks can be created, for example, by using alternate operations or alternate numbers or types of numbers (e.g., three-digit numbers instead of two-digit numbers). A form such as the one shown here can serve as a convenient template for creating your own open questions and parallel tasks. The Appendix includes a full-size blank form and tips for using it to design customized teaching materials.

Geometry

DIFFERENTIATED LEARNING activities in geometry are derived from applying the NCTM process standards of problem solving, reasoning and proof, communicating, connecting, and representing to content goals of the NCTM Geometry Standard, including

- analyzing characteristics and properties of two-dimensional (2-D) and three-dimensional (3-D) shapes, and developing mathematical arguments about geometric relationships
- specifying locations and describing spatial relationships
- applying transformations and using symmetry to analyze mathematical situations
- using visualization, spatial reasoning, and geometric modeling to solve problems (NCTM, 2000)

TOPICS

Students move from working with shapes by focusing on overall appearance to focusing on particular characteristics and properties (Small, 2007). They first make sense of shapes by handling and viewing the actual shapes and only later by using pictorial representations. As students develop, they become increasingly able to transform shapes and to make informal deductions. The NCTM Curriculum Focal Points (NCTM, 2006) is a useful source for seeing how instruction in geometry is typically sequenced.

Prekindergarten–Grade 2

Within this grade band, students begin by informally describing the shapes around them and using **positional vocabulary** to more specifically name 2-D and 3-D shapes and describe location. They begin to build structures and observe how large shapes or structures can be made up of smaller shapes; then they more specifically describe and compare shapes and learn to recognize them from different perspectives. By combining shapes, students are building a foundation for later work on area and with fractions.

Grades 3–5

Within this grade band, students begin to use specific **properties** of 2-D and 3-D shapes—such as side relationships, angle relationships, or types of **faces**—to classify them. They build, represent, and analyze both 2-D and 3-D shapes to better understand them. Students specifically consider **congruence** and both **line symmetry** and **rotational symmetry** in 2-D shapes, as well as the number of faces, **vertices**, and **edges** of 3-D shapes. They explore how shapes can be combined and dissected to support later work with area and fractions in two dimensions and volume in three dimensions.

Students also begin an exploration of **transformations** as strategies to design and analyze changes of positions of shapes in two dimensions.

Grades 6–8

Within this grade band, students focus on decomposing shapes in two dimensions and three dimensions to support work in area and volume. They also solve more complex geometry problems than younger students would, including locating shapes on **coordinate grids** and using the **Pythagorean theorem**, which relates lengths of sides in a right triangle.

Another topic at this level is **similarity,** the result of enlarging or reducing a shape so that the overall proportions, but not necessarily the size, remain the same.

THE BIG IDEAS FOR GEOMETRY

Coherent curricula in geometry that meet NCTM content and process standards (NCTM, 2000) and support differentiated instruction can be structured around the following big ideas:

- Shapes of different dimensions and their properties can be described mathematically.
- There are always many representations of a given shape.
- New shapes can be created by either combining or dissecting existing shapes.
- Shapes can be located in space and relocated by using mathematical processes.

The tasks set out and the questions asked while teaching geometry should be developed to reinforce these ideas. The following sections present numerous examples of application of open questions and parallel tasks in development of differentiated instruction in these big ideas across three grade bands.

OPEN QUESTIONS FOR PREKINDERGARTEN–GRADE 2

OPEN QUESTIONS are broad-based questions that invite meaningful responses from students at many developmental levels.

✳ **BIG IDEA.** **Shapes of different dimensions and their properties can be described mathematically.**

> You are making a book about yourself. Draw a shape for the cover of your book that tells something about you. Why did you choose that shape?

A question such as this one provides the opportunity to explore what students know about properties of shapes. As they talk about why they chose the shape they did, students may reveal a lot about themselves (an added bonus), but it is possible to direct the conversation toward geometric **attributes** of shapes by reminding students to describe what aspects of the shape make the selection sensible.

The question is suitable for a broad range of students because students who need to can choose a simple shape so that they will be able to explain their choice effectively.

> Use shape stickers to make a shape picture. Try to use as many different shapes as you can. Describe your picture and the shapes in it. Tell why you used those shapes.

A task such as this one provides the opportunity to see what students know about shapes, and it allows the teacher to see how students combine shapes to make other shapes. As students talk about why they chose the shapes they did, the teacher can elicit their understanding of where various shapes appear in their environment and build on that.

Again, students have the option of using simple shapes if that is all they are comfortable with, or they can use very complex shapes if they wish to.

Variations. It is possible to reuse this task by adding stipulations, for example, by indicating what shapes can be used, how many shapes can be used, what the picture must be about, and so on.

Open Questions for Pre-K–Grade 2

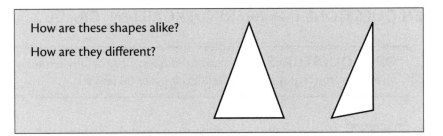

How are these shapes alike?

How are they different?

This question is open in that students can choose any aspects of the shape to focus on. When they make their choice, it becomes clear what they do or do not know about triangles. Some may talk about symmetry or lack of symmetry; some will use geometry vocabulary, for example, *They are both triangles.* Others will notice that both shapes have three sides without using the term *triangle.* Some students may even think about how the shapes can be divided or what other shape each one is a part of.

Variations. It is possible to reuse this question with other pairs of shapes: for example, a triangle can be paired with a **quadrilateral**, a 2-D shape with a 3-D shape, or two different 3-D shapes (e.g., a cube and a long thin **rectangular prism**) can be compared.

TEACHING TIP. Some questions can be varied quite simply to be used over and over again. The question here is just one example.

Choose a type of shape. Tell as many things about it as you can.

This very open question allows students to tell whatever they know about a shape, whether it is 2-D or 3-D. Students who describe a shape reveal what they understand about the attributes or properties of the shape. Students should be encouraged to focus on geometric attributes rather than other attributes such as color or texture if the discussion begins to shift in that direction.

A student who picks a circle might say that it is round, that it is curved, or that there are no ends to it. A teacher can build on what the student has said by asking further questions. For example, the teacher might ask:

- *Can you think of a shape that does have ends?*
- *What do the ends look like?*

TEACHING TIP. One of the easiest ways to frame an open question is to mention an object, a property, a number, a measurement unit, a type of pattern, a type of graph, and so on, and ask students to tell everything they know about it.

Open Questions for Pre-K–Grade 2

✶ **BIG IDEA.** New shapes can be created by either combining or dissecting existing shapes.

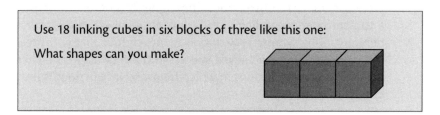

> Use 18 linking cubes in six blocks of three like this one:
>
> What shapes can you make?

Often geometry instruction focuses on 2-D geometry rather than 3-D geometry because that is more convenient in a classroom where there are many students but relatively few manipulatives available. However, instruction in 3-D geometry is extremely important to help students develop spatial sense. Questions such as the one above require only simple materials and are useful for geometric development.

In this case, it will be interesting for the teacher to see whether some students create only flat shapes (one layer high) or think only of traditional shapes, such as **prisms**. Students who are ready might use problem-solving strategies such as an organized list to come up with more options.

Variations. Similar questions can be created by changing the number of small structures being combined, as well as the number of cubes or arrangement of cubes within the small structures.

> Draw a design or shape made up of three shapes. The design should have symmetry.

Symmetry is a geometric property that is very important for students to understand because it is evident in the everyday world. This particular task is open in that students can use simple or more complex shapes, as they choose. No matter what choice students make, they will still be focusing on the concept of symmetry.

As students discuss their shapes, each student in the class has the opportunity to contribute, because his or her design or shape is likely to be different from every other student's.

Variations. Variations of the question can be created by changing the conditions. For example, the number of shapes might not be three, or stipulations might be added about which shapes can be used.

TEACHING TIP. Students do not always follow instructions perfectly. Rather than losing sight of an accomplishment in favor of achieving complete adherence to instructions, sometimes it is better to focus instead on a student's understanding of the concepts involved. In this case, a student demonstrating symmetry with two shapes or four shapes (instead of the requested three) has achieved success in the geometry task. There will, of course, be other situations in which following instructions is much more critical.

> Begin with a shape of your choice. Cut it into three pieces. Give it to a partner to put together.

Not only should students have experience putting together shapes to make new ones, they should also have experience dissecting shapes into other shapes. These skills will support later work with fractions, area, and volume, and they will also help students make sense of more complex shapes they might meet.

In this situation, a student might start with, for example, a square and see that it can be cut into three pieces that are familiar shapes or three pieces that may be more unusual. Students might be surprised to learn that sometimes it is easier to put the pieces back together if the shapes are unusual (how pieces fit together is sometimes more obvious in these cases).

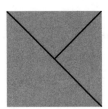  

This task may be less intimidating to some students than other tasks would be because they have the freedom to cut wherever they wish.

✸ `BIG IDEA.` **Shapes can be located in space and relocated by using mathematical processes.**

> Choose two objects in the room. Think about their locations. Tell how to get from one location to the other.

Position is an important aspect of spatial thinking. Even young students can develop very simple spatial maps. Open questions such as the one above are valuable because a struggling student has the option of selecting two objects whose positions rest on the same horizontal or vertical line, thus simplifying the task of locating the objects in relation to one another.

The description of the path to get from one location to the other can be verbal or pictorial. A pictorial representation will reveal whether students have considered measurement concepts at all. For example, if the student has to go north–south a much greater distance than east–west, does the representation reflect that?

Variations. Additional questions can be created by asking students to describe paths that include more objects.

OPEN QUESTIONS FOR GRADES 3–5

✹ **BIG IDEA.** Shapes of different dimensions and their properties can be described mathematically.

> A certain shape makes you think of a rectangle, but it is not a rectangle. What could it be? Why?

There are many possible responses students might come up with. A few are shown here:

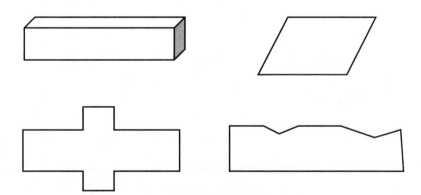

Because the question asks for a shape that makes the student think of a rectangle, there is no risk in responding—anything would be acceptable as long as students offer a reason as to why they see a connection. But the question does allow for a rich discussion of shapes in which all students can take part. Students could discuss the relationship between 2-D and 3-D shapes if a student chooses, for example, a rectangular prism. Or a student might bring out the relationship between **parallelograms** and rectangles (affording the opportunity for the teacher to explain that a rectangle is a parallelogram). Ideas of symmetry might be discussed if a student chooses a shape with the same kind of symmetry a rectangle has.

Variations. The question can be varied by asking about other connections, for example, shapes that make the student think of a circle but are not circles.

TEACHING TIP. Differentiated instruction is supported when the environment encourages risk taking. It is only when students really feel that their teacher is open to many ideas that they start to focus on the mathematics rather than spending their time guessing what it is their teacher wants to hear.

> A triangle is made by using **Cuisenaire rods** or centimeter strips. The **perimeter** is 10. How many different triangles are there? What are their side lengths?

Often, but not always, geometry and measurement concepts are intertwined. Although this question involves measurement (perimeter), the focus is actually on the geometric concept that the two short sides of a triangle must, in total, be longer than the longest side. For example, 8, 1, 1 cannot be the sides of a triangle because the two 1 sides cannot meet and still both touch the 8 side.

The question proposed is more suitable for struggling students if the rods are made available. A triangle with a perimeter of 10 (i.e., three sides of length totaling 10) is a fairly simple one to create. Students could use an orange 10 rod and begin by using a variety of triples of rods to make up that total length and then try to make a triangle with the three rods selected.

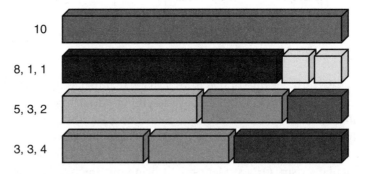

The mathematical differences among students will become evident in their ability to generalize to the concept that some combinations are not even worth trying (e.g., one long and two very short sides) and in their ability to decide whether they have considered all possible combinations.

Variations. One way to vary this question is to not stipulate that rods be used (thus, the length limit is no longer 10). Instead, centimeter strips of up to 15 cm long might be used. Alternatively, students can be asked to create quadrilaterals, or additional conditions can be placed on the problem, for example, the triangle shape might have to be **isosceles**.

> Build a shape with nine linking cubes. It has to be at least two cubes wide and at least two cubes tall. It also has to be symmetric.

This question brings to the forefront the important concept of symmetry. However, because it focuses on 3-D shapes, it extends symmetry into three dimensions. This topic often gets very little attention in school, but it has many practical applications for students because they live in a 3-D world.

Students are allowed choice in the shape they build but with some constraints. They can choose to construct simple shapes, if that is all they are comfortable with, or more complex shapes. By stipulating use of nine cubes, the question will bring out the notion that the center of the shape must incorporate an odd number of cubes because an even number is required to make the sides symmetric.

Two possible shapes are shown below:

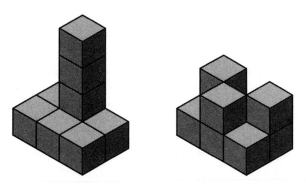

Variations. By varying the conditions, for example, changing nine to a different number, or changing the conditions on width and depth, other problems can be created.

> Make a square on a **geoboard**. Move the elastic so that one corner is no longer a corner. What different shapes can you make?

By beginning with a square, every student can participate. Stronger students will realize that by starting with a larger square, they have many more options in terms of the look of shapes they can create. Because there are so many possible results, the discussion of the question will likely be a rich one.

Open Questions for Grades 3–5

Some of the options are shown below:

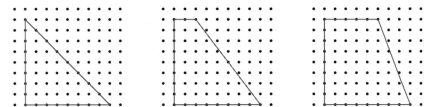

Variations. The question can be altered to create additional questions by allowing the student to begin with other shapes and to remove more than one corner, if desired.

> What shapes can you build with two identical short straws and two identical long straws?

Students can be given two short and two long straws. Some of them will consider only **polygons**, whereas others may allow for other sorts of shapes as well.

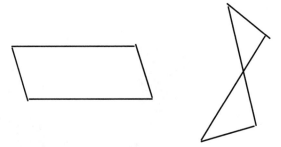

Some students will investigate what they can do only with the given straw lengths. For example, given two 10-cm and two 12-cm straws, they could make a parallelogram of specific dimensions. However, other students might generalize to what happens with any two pairs of side lengths. Some students will assume that all four straws must be used, and others will consider the possibility of using only three of the straws or maybe even only two of them (to create angles). Every student in the class will be able to attack the problem at some level.

Variations. The question can be varied by changing the number of straws or the numbers of pairs of straws that are congruent.

✳ **BIG IDEA.** There are always many representations of a given shape.

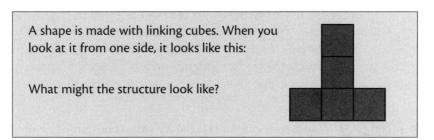

A shape is made with linking cubes. When you look at it from one side, it looks like this:

What might the structure look like?

This task is an open one in that students are not told how many cubes to use or what the shape might look like from other perspectives. This allows all students to find an appropriate entry point. Some students will assume that the structure has only the five displayed cubes in it, whereas others will realize that there can be other cubes as well:

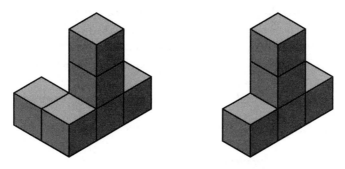

One way to vary the complexity of this question is to decide whether or not to provide linking cubes for the students to manipulate. More advanced students might be asked to visualize without actually having the cubes, although most students would benefit by having the concrete materials available.

The question can be made more complex by telling students how many linking cubes are used. If there are no conditions stated, as above, the question is much simpler than if, for example, it is stipulated that there are 11 cubes in the structure.

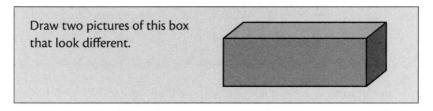

Draw two pictures of this box that look different.

Students will, of course, have varied drawing abilities, but mathematically the focus of the question is on how students might represent the shape. The fact that the shape must be drawn in two different ways will ensure that students consider transformations, for example, what the shape will look like if it is turned or viewed from above.

Some students might draw only the front face; others might draw all six faces. Still others might try to draw a sketch that makes the shape look 3-D or might draw only the **skeleton**. Observing how students consider the representations of the shape will provide the teacher insight into what aspects of the shape are clear to students and what aspects need to be brought out in further instruction.

✹ BIG IDEA. **New shapes can be created by either combining or dissecting existing shapes.**

> How many different shapes can you make by using five green **pattern block triangles**? Triangles must match along full sides.

By providing students with the pattern blocks and allowing them to experiment, the problem is made accessible to all students. Some students will use more sophisticated organizational skills to make sure they have all the possibilities; others will be more random in their approach to the problem. However, all students can contribute to a discussion of what they discovered.

The problem also provides an opportunity for discussion of congruence. For example, students might debate whether or not these two shapes are really different:

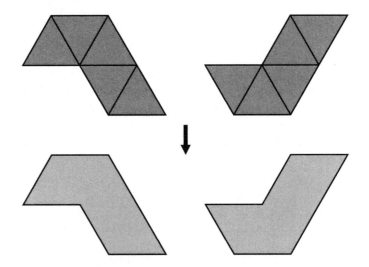

Variations. Rather than providing enough green blocks for students to make many shapes, it is possible to make the original question more challenging by providing only five green blocks to each student, along with recording paper. Then students must take the extra step of going from the concrete to the pictorial to keep track of what they have done.

TEACHING TIP. When handling discussion in a whole group setting, the teacher should call on some of the weaker students first to make sure that all of their ideas have not already been stated by the time they are called on.

Here is one **tangram** animal.

Make your own tangram animal.

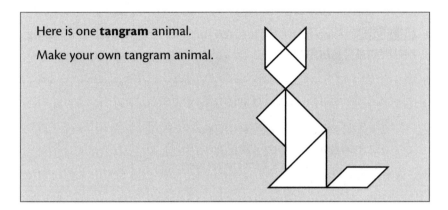

To answer this question, students would need to have been introduced to the tangram puzzle in advance. The tangram is a set of seven pieces that make a square, as shown below:

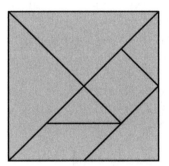

By providing the animal example, the question allows even the weaker student to succeed by copying the animal given. However, by opening up the task to allow students to create their own animals, all of them can be appropriately challenged. Tangrams are widely used, and there are many sources for other ideas for animals, as well as for numerals, letters of the alphabet, and so on (Tompert, 1990).

Use any four tangram pieces to build a shape that looks like a house.
Use geometry words to describe your house.

By not showing the shape the student must achieve, the question is much more open. A student can claim that his or her shape is like a house no matter what it looks like and be correct. The important part is for the student to use geometry language when he or she describes the shape. The language the student uses can be as simple or as complex as suits the individual. For example, some students will name their final shape (e.g., as a heptagon, a shape with seven sides), whereas others will simply describe the shapes that make up their house (e.g., a square and three triangles).

✳ **BIG IDEA.** **Shapes can be located in space and relocated by using mathematical processes.**

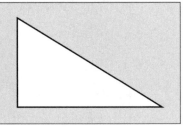

You start with the triangle shown here:

You flip the triangle by using a **transparent mirror**. It moved a little, but not a lot. Where was your **flip line**?

By using language like "a little, but not a lot," there is additional latitude for students to put their personal stamp on the problem. In this case, a student might suggest that the flip line is the **hypotenuse** of the right triangle (although many would not use that term); some of the triangle would not have moved at all, specifically the hypotenuse, and so the triangle would be viewed as not having moved a lot. Other students might do a flip through the triangle, and its overall position would move even less.

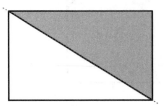 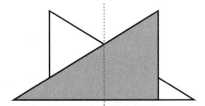

Whatever solutions students come to—and there are many reasonable solutions—the goal of the activity will have been achieved: consideration of how reflection affects a shape.

TEACHING TIP. Often an ambiguous term such as "a lot" or "a little" makes a mathematical question much less intimidating to students than using a specific value.

OPEN QUESTIONS FOR GRADES 6–8

✷ **BIG IDEA.** Shapes of different dimensions and their properties can be described mathematically.

> We use different words to describe triangles. For example, we might call them **acute**, **obtuse**, **right**, **equilateral**, isosceles, or **scalene**.
>
> What combinations of two words can be used to describe a triangle?

Because the question asks for which combinations are possible, and does not insist on all possible combinations, the question is appropriate for struggling learners as well as students with more developed understanding. For example, a struggling student might simply draw a triangle, notice that it is both right and isosceles and list this single combination as a possibility. Other students may want to consider all possible combinations.

In follow-up questions, students can be asked which combinations are not possible and why.

> A shape has six sides and two 90° angles. What could it look like?

Some possible shapes are shown below:

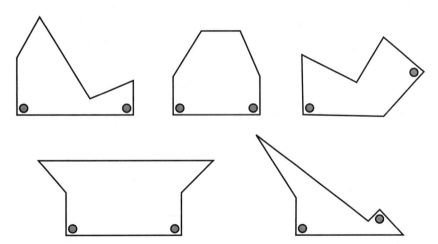

Some students will focus on starting with a line segment with two right angles at either end, then making sure there are six sides altogether, whereas others might recognize that the **right angles** do not need to be attached to the same side of the shape. Again, the question provides students the opportunity to offer more and less standard responses.

The teacher could follow up by asking what would happen if there were only four sides. Some students will recognize that the types of shapes then become much more limited:

Variations. The question can be varied by changing the conditions on the number of sides, the number of angles given, or the sizes of the angles given.

> You are working on a 5 × 5 dot grid. You use dots on the grid as vertices of polygons. What polygons with a lot of sides can you create?

By posing the question using the phrase "a lot of sides," rather than, for example, the maximum number of sides, the question is open to more students. When asking for the maximum, there is a single right answer. With this open question, a student who comes up with, say, 15 sides can legitimately be seen to be successful even though someone else may come up with more sides.

Variations. Similar questions can be asked on different-sized grids.

> The **diagonals** of a quadrilateral meet, and one of the angles formed is very small. What could the quadrilateral look like?

As with some of the other questions and tasks posed, this question is open enough to allow for a student to consider either a single possibility or many possibilities. A student might draw any or all of the shapes shown:

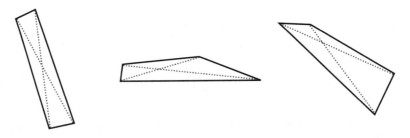

The question provides the opportunity for teachers to follow up by asking students what they notice about the three other angles formed where the diagonals cross. They are likely to notice that there are always two pairs of equal angles, and that if one pair of angles is small, the other pair must be large.

✳ **BIG IDEA.** There are always many representations of a given shape.

> A shape has a triangle for a **cross-section**. What could the shape be?

This question is suitable once students know what a cross-section is. Some students will assume that a **pyramid** is the logical shape to try and will find that it is not as easy as they thought to get a triangle cross-section. It is possible, for example, if a **triangle-based pyramid** is cut parallel to a base or at a vertex.

Other students will recognize that if a cut is made whenever three faces meet at a vertex, a triangle can be formed. Thus, a cube could be cut to create a triangular cross-section:

The question is suitable for a broader group of students if modeling clay is provided for students to make shapes and dental floss to allow them to cut cross-sections.

✳ **BIG IDEA.** New shapes can be created by either combining or dissecting existing shapes.

> What can you make if you put together three identical **isosceles trapezoids**?

Some of the possible answers are shown below:

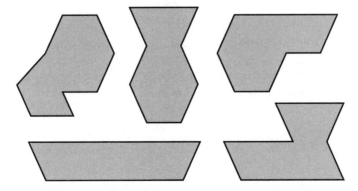

If students are provided with isosceles trapezoids, such as the red pattern blocks, they can experiment concretely. If they are given many copies of the original shape, they can simply put them together and leave the results on their desks. If they are given only three **trapezoids**, they need to focus on how to make a copy of what they have done (by tracing or drawing) before they go on to the next shape. This is more suitable for some students than for others.

Some students will think of more possibilities and more complex shapes than others, but almost any student can come up with at least one shape.

Variations. The question can be varied by changing the type of shape that is used as the basic unit and the number of copies of the shape that are required to make the larger shapes.

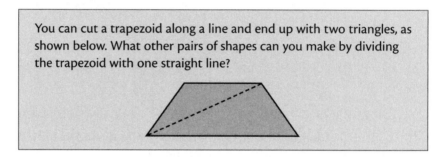

You can cut a trapezoid along a line and end up with two triangles, as shown below. What other pairs of shapes can you make by dividing the trapezoid with one straight line?

As with other questions with many possible answers, many students have the opportunity to contribute to a discussion about this question. Weaker students might simply divide the trapezoid into two different triangles by using the other diagonal, an acceptable response. Other students will think of other possibilities, for example, a right triangle and a **right trapezoid**, an isosceles triangle and a parallelogram, or two trapezoids.

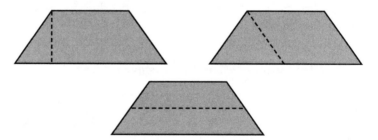

Variations. The question can be adapted by changing the starting shape or by changing the number of dividing lines.

Show how to put together squares to create shapes with eight sides.

Again, some students will consider only one possibility, whereas others will realize that different numbers of squares can be put together to make shapes with eight sides. Probably all students will realize that at least some of the squares used must be staggered in order to increase the number of sides from four to eight.

Some possible solutions are shown.

Using only two squares:

Using three squares:

Using four squares:

Using five squares:

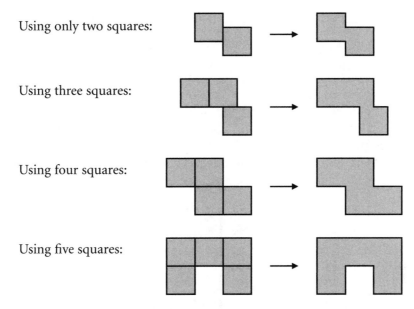

Variations. The question can be varied by allowing students to start with shapes other than squares or by asking them to create shapes with a different number of sides.

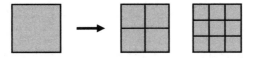

Can you usually divide a shape up into one or more similar shapes?

To approach this problem, students will need to know that shapes that are similar are enlargements of or reductions of (or identical to) a given shape. A student is likely to begin with a simple shape such as a circle and try to divide it up. They will see that it is not possible to divide up a circle into only circles that are smaller.

If they try a square, they will see that it can be divided into more squares, in more than one way:

Some students will quit there, but others will investigate further, to see for which types of shapes this kind of division is possible and for which it is not. Because a simple shape like a square provides a solution, the problem should be accessible to all students in the class.

✳ **BIG IDEA.** **Shapes can be located in space and relocated by using mathematical processes.**

> One vertex of a triangle is at the point (1,2). After a **reflection**, one vertex is at the point (5,8). Name all three vertices of the original and final triangles.

Some students will only be comfortable with either horizontal or vertical **reflection lines**. The way the question is posed allows them success even with this limitation.

For example, the original triangle could have been positioned with vertices at (1,2), (1,8), and (3,8) and the new triangle with vertices at (5,2), (5,8), and (3,8):

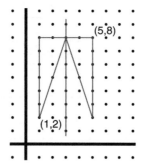

Other students will assume that it was the point (1,2) that moved to (5,8). They will look for a "diagonal" reflection line that accomplishes this task. In fact, they could use such a line and the original vertices (1,2), (3,5), and (6,3) move to (5,8), (3,5), and (6,3):

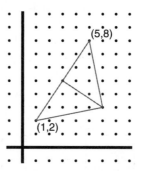

Still other students will recognize that there are even more possibilities than that. For example, the original triangle could have been positioned with vertices at (1,2), (0,3), and (−2,1) that move to (5,8), (6,7), and (7,10).

A shape is completely in the first **quadrant**. Where could it be after a **translation**?

Some students will try one shape and one translation and come up with an answer. For example, a student might draw a square and move it up or to the right and correctly point out that the shape has stayed in the first quadrant. Other students will realize that they could translate sufficiently to the left to end up in the second quadrant, sufficiently down to end up in the fourth quadrant, or sufficiently left and down to end up in the third quadrant. Still other students will look for all the possible combinations, showing that the shape could end up in one, two, three, or even all four quadrants.

Variations. Similar questions can be created by changing the transformation (to rotations, reflections, or even dilatations) or by stipulating the shape with which to start.

PARALLEL TASKS FOR PREKINDERGARTEN–GRADE 2

PARALLEL TASKS are sets of two or more related tasks that explore the same big idea but are designed to suit the needs of students at different developmental levels. The tasks are similar enough in context that all students can participate fully in a single follow-up discussion.

✸ **BIG IDEA.** Shapes of different dimensions and their properties can be described mathematically.

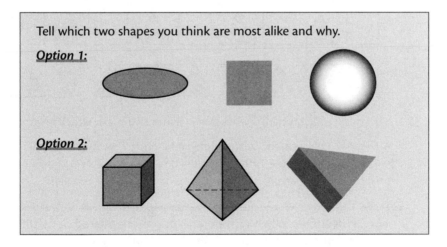

Tell which two shapes you think are most alike and why.

Option 1:

Option 2:

In this set of tasks, the choice is less about the developmental levels at which students are operating than it is about their comfort with a broad variety of shapes.

Some students may prefer *__Option 2__* because they see a similarity in the shapes: all are **polyhedra** (they may not use that word, of course), or shapes with faces that are all polygons. They might focus on the fact that the second and third shapes both have triangular faces or that the first and third shapes both have rectangular faces.

On the other hand, some students might prefer *__Option 1__* because they see two rounded shapes. Other students might see the square and oval as more similar because both are flat.

In follow-up discussion, regardless of which option was selected, in addition to asking which shapes the student chose and why, the teacher could ask for additional observations from classmates, leading to a lively exchange to which students of all levels could contribute.

TEACHING TIP. Observing the choice a student makes when a choice is offered provides important diagnostic information to the teacher.

> *__Option 1__*: Use your pattern blocks. Create a shape pattern where the 10th shape is a green triangle.
>
> *__Option 2__*: Use your pattern blocks. Create a shape pattern.

The distinction between these two options is the constraint that is imposed within *__Option 1__*. This constraint might be seen by the student as increasing the complexity of the problem. No matter which task students complete, they are required to think about the properties of shapes and their position or orientation to be able to form patterns.

Two patterns that would satisfy either option are shown below:

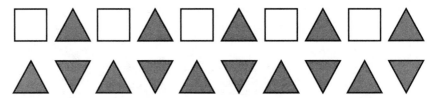

In discussing students' work, the teacher could ask either group of students:

- *Describe your pattern. What could come next? How do you know?*
- *What shape is in the 10th spot? How do you know?*
- *How could you create a different pattern that fits the problem?*

TEACHING TIP. When numbered options are offered, sometimes the "simpler" option should be presented as Option 1 and other times as Option 2. The unpredictability will ensure that students consider both possibilities when they choose their tasks.

> **Option 1:** Use 20 toothpicks to make three shapes. None of the shapes can use the same number of toothpicks. Describe your shapes.
>
> **Option 2:** Use 20 toothpicks to make at least four shapes. Describe your shapes.

Students choosing **Option 1** might create shapes with many sides and only a single toothpick on each side, or shapes with more than one toothpick making up a side, or some combination of both approaches. For example, two possible solutions are:

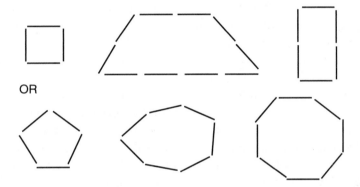

OR

Some students might prefer **Option 2**, where they could use familiar shapes more than once and need not ever create shapes with many sides.

In follow-up discussion, all students could share their thinking no matter which option they chose.

> **Option 1:** Choose 2-D shapes to make two different creatures. Describe the two creatures you made.
>
> **Option 2:** Choose 3-D shapes to make two different creatures. Describe the two creatures you made.

By providing the option of working with 2-D or 3-D shapes, the problem becomes accessible to more students. Because there are no rules for what a creature looks like, there is also great freedom in which shapes are used and how they are used.

No matter which task the students chose, the teacher could ask:

- *What are the names of the shapes you used?*
- *How many of each did you use?*
- *Why did you decide those would be good shapes to use?*
- *How did you decide which shapes would be next to which other ones?*

✸ **BIG IDEA.** **There are always many representations of a given shape.**

> Pick one of these shapes. Represent it as many ways as you can. Use pictures, words, numbers, and objects.
>
> cube   square   cylinder   **hexagon**

In posing these parallel tasks, the teacher should have a variety of materials available, including molding clay, drawing materials, paper that can be cut or rolled, pattern blocks that can be used or traced, 3-D models, straws and connectors, and **polydrons** (plastic shapes that can be connected to form 3-D shapes). These can be supplemented with other found items in the classroom; for example, a student might use a book with a front cover in the shape of a square as a model for the square. The choice in this task allows students more comfortable with 2-D to work in that realm, whereas other students can choose 3-D shapes.

A student who selects the cube might simply find models of cubes in the classroom (e.g., dice), might draw a picture of a cube, or might draw all of the individual faces of a cube. A student who chooses the hexagon may not initially realize that there are many types of hexagons—not all of them the standard **regular hexagon**—but may come to recognize that fact as he or she searches for six-sided objects. A student who chooses the cylinder might roll a piece of paper to make a cylinder, thereby setting up a future connection that will be useful in exploring the surface area of cylinders.

> _**Option 1:**_ Use your shapes to make a "map" of our classroom that shows where my desk, your desk, and the door are.
>
> _**Option 2:**_ Draw a "map" of our classroom that shows where my desk, your desk, the blackboard, the reading corner, and the door are.

These choices allow students to use either concrete shapes or to draw pictures, whichever is more comfortable for them. The choices acknowledge the fact that students might be at different developmental levels because one of the distinctions between students' geometric ability is their comfort in using pictorial representations rather than concrete representations. In addition, _**Option 2**_ requires the student to include a few more items in the map.

Students can share their maps no matter which option they complete. They can describe which object they placed first and how they placed subsequent objects. They can explain which shapes they chose to represent the objects in the room and why. Students might even be asked how they would describe a path from one object to another on the map.

✳ **BIG IDEA.** New shapes can be created by either combining or dissecting existing shapes.

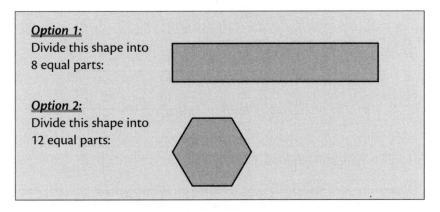

Many students will find it easier to divide the rectangle than the hexagon; they can think about folding the rectangle, but might find folding the hexagon more complex. On the other hand, the hexagon task is not overly difficult because the student might realize that he or she could first create the six triangles that make up the hexagon and then divide each one. Recognizing this possibility is more likely if students have had prior experience working with pattern blocks.

Both tasks focus on the idea that a shape can be divided into many smaller shapes. The follow-up for either task could involve asking students to talk about alternate ways to divide the shape.

Variations. The tasks can be adapted by using other shapes and other numbers of parts.

PARALLEL TASKS FOR GRADES 3–5

✳ **BIG IDEA.** Shapes of different dimensions and their properties can be described mathematically.

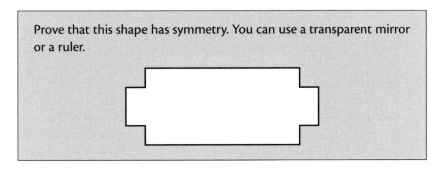

Research has indicated that using a ruler to test for symmetry is a more developmentally demanding task than using a transparent mirror (Small, 2007). Using a ruler requires students to understand that the **perpendicular distance** to any point on one side of the reflection line is equal to the distance to its matching point on the other side.

By allowing students their choice of tool, the needs of students at different developmental levels are being considered.

TEACHING TIP. If students who should be working at a more challenging level consistently opt for the simpler task, the teacher should suggest to individual students or individual groups of students which tool they should use.

✴ **BIG IDEA.** There are always many representations of a given shape.

> *Option 1:*   A prism and pyramid have the same number of corners. Draw what they could look like or build them.
>
> *Option 2:*   A prism and pyramid have the same number of edges. Draw what they could look like or build them.

Providing students with small balls of clay or marshmallows to serve as the corners, and straws or toothpicks to serve as the edges, enables them to model the shapes by building skeletons.

Students who choose **_Option 1_** will likely build the prism first. Then, once they realize that there is only one corner at the top of the pyramid, they will know exactly how many corners the base must have. For example, if a student builds a **triangle-based prism** (which has six corners), he or she will realize that a **pentagon** must form the base for the pyramid. If students start with the pyramid first, the problem may be more challenging unless they happen to realize that it is essential to start with an even number of corners on the pyramid.

Students who choose **_Option 2_** may begin with the prism or the pyramid. Because it turns out that the number of edges on a prism is always triple the number on the base and the number of edges on a pyramid is always double the number on the base, students will discover that the problem can only be solved if the number of edges used is a multiple of both 3 and 2.

Both of these problems are challenging but manageable. The second is clearly more complex than the first.

Parallel Tasks for Grades 3–5

✸ **BIG IDEA.** New shapes can be created by either combining or dissecting existing shapes.

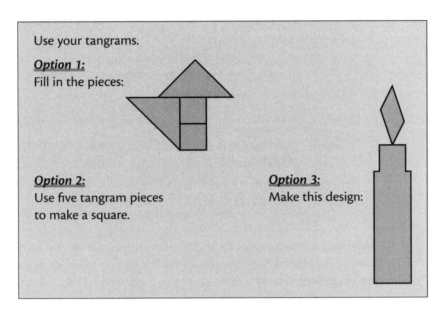

Use your tangrams.

Option 1:
Fill in the pieces:

Option 2:
Use five tangram pieces
to make a square.

Option 3:
Make this design:

Parallel Tasks for Grades 3–5

Option 1 is most suitable for students whose geometric concepts are less highly developed. If the diagram is made life size, students can simply experiment and try different pieces in different spots. In discussion at the conclusion of the work, those students are able to talk about how they fit the shapes in, what clues they used, and so on.

Option 3 is a mathematically more sophisticated activity than **Option 1** because no internal lines are presented. Its complexity, too, can be varied based on whether the diagram is life size. A life-size diagram will always be easier to work with because students will be able to fit pieces into the outline. The diagram at the right shows the tangram candle with outlines of the component pieces.

Option 2 is the most open. An even more open alternative would be to simply ask students to use any number of tangram pieces to make a square, offering some very simple solutions. By requiring exactly five pieces, the task poses more challenge to students than it would in its simplest form. The diagram at the right shows a square made up of five tangram pieces, with outlines.

The follow-up discussion could invite students, no matter what task they chose, to describe how they went about solving their problem.

Variations. **Option 1** can be adjusted to be slightly more difficult by using a life-size diagram with no internal lines shown.

✹ **BIG IDEA.** **Shapes can be located in space and relocated by using mathematical processes.**

> Use this triangle:
>
> *Option 1:* Make two new shapes by **sliding** and flipping the triangle.
>
> *Option 2:* Make two new shapes by flipping and turning the triangle.
>
> *Option 3:* Use three copies of the triangle. Make two new shapes by putting the copies together in different ways.

Rotations, or turns, seem to be the most difficult transformation for students. By allowing students the option of using only slides and flips, or of simply combining shapes, more students will be able to succeed at the task they choose.

Some possible solutions are:

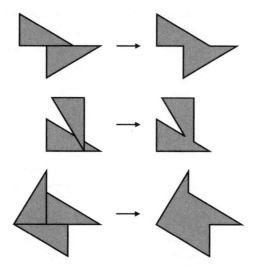

In the follow-up discussion, questions such as these could be asked of all students:

- *What do your new shapes look like?*
- *How did you create your new shapes from the original one?*
- *Could you have done your steps in a different order and ended up with the same shape?*

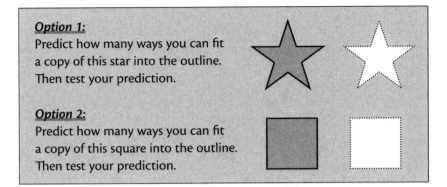

Option 1:
Predict how many ways you can fit a copy of this star into the outline. Then test your prediction.

Option 2:
Predict how many ways you can fit a copy of this square into the outline. Then test your prediction.

Each of these tasks involves considering both rotational and line symmetry. Because the square is more familiar, many students will be more comfortable with that choice. On the other hand, some students will choose the star because it is a bit more obvious that pointy sections must fit into pointy sections to make the copy fit into the outline.

Some students will consider only rotational symmetry, some only line symmetry, and some both. In the course of discussing the results of their work, students who have considered only one type of symmetry will hear about the other type from other students.

In the follow-up conversation, all students could be asked to describe:

- *How they went from one possible solution to the next one.*
- *How they knew when they had all the possibilities.*
- *Why they chose the shape they did.*

PARALLEL TASKS FOR GRADES 6–8

✴ **BIG IDEA.** Shapes of different dimensions and their properties can be described mathematically.

These two dots are the corners of a given shape:

Option 1: Make a square with these corners.

Option 2: Make a parallelogram with these corners.

Option 3: Make an isosceles triangle with these corners.

Some students will favor *Option 3* because they only need to place one more point. The fact that the triangle must be isosceles makes the task appropriately challenging, however.

Some students who choose **_Option 1_** will choose to make the dots the end-points of a diagonal so that the sides can be horizontal and vertical. Others will realize that the square can be on a slant and will use the dots as the endpoints of one side.

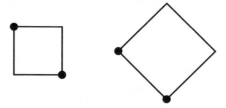

Still other students will see the slant inherent in the segment joining the two points and use horizontal lines to make the parallelogram specified in **_Option 2_**.

No matter which task was selected, the student could be asked:

- *Describe your shape.*
- *What name would you give your shape? Is it a … [e.g., a polygon]?*
- *How do you know it is the shape you say it is?*

Option 1: Draw a shape where one side is 4" long and one angle is 30°.

Option 2: Draw a hexagon where one side is 4" long, one side is 6" long, one angle is 30°, and one angle is 150°.

These two tasks are similar in that both set out conditions for drawing a shape. **_Option 1_** is much less restrictive and may be more suitable for some students.

The student who chooses **_Option 1_** merely needs to draw a 4" length and put an angle of 30° at one end of it. The shape drawn is up to the student, so he or she might draw a simple triangle or something more complex. The student who chooses **_Option 2_** must ensure that the shape is a hexagon and that all four constraints on the hexagon are considered.

Discussion of the results of the exercise would allow students to describe how they created the shape—what they did first, then next, and so on. The teacher could ask why they chose the particular first step they did rather than a different first step. It would not matter which option the the students chose; the same questions would be relevant for all tasks.

Variations. **_Option 1_** can be modified by adding one constraint, for example, requiring that a hexagon is created. This more complex task can be offered as a third option to accompany the other two, or as a replacement for the original **_Option 1_**.

✺ **BIG IDEA.** There are always many representations of a given shape.

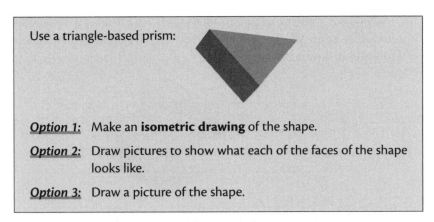

Use a triangle-based prism:

Option 1: Make an **isometric drawing** of the shape.

Option 2: Draw pictures to show what each of the faces of the shape looks like.

Option 3: Draw a picture of the shape.

Research has shown that drawing a picture of each face is easier than creating an isometric drawing (a drawing on a triangular grid that shows equal lengths on a 3-D object as equal on the 2-D representation) (Small, 2007). Drawing a rough picture of the full shape is generally easier than making sure that each face is represented. By setting out these three alternatives, students can work on the task that is most suitable for them.

Again, no matter which option students complete, they could be asked to share their drawings, tell how they started and then continued the task, and tell how confident they are that they have a good picture of the shape.

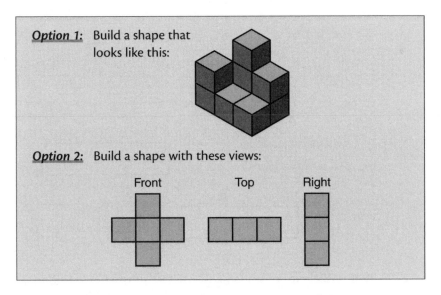

Option 1: Build a shape that looks like this:

Option 2: Build a shape with these views:

Front   Top   Right

Both tasks allow students to consider 2-D representations of 3-D shapes. The first option allows students to use simpler visualization skills than the second one, but both require them to analyze the presented information and translate what they see into another form.

For example, in **_Option 1_**, students must consider what cubes might be hiding behind those shown and whether or not to include them in the shape they build. In **_Option 2_**, students must consider which view to start with and then ensure that the shape constructed will produce all three required views. The elegantly simple solution is shown below:

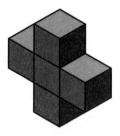

In discussing their work, all students can talk about how they created the shape, what their shape looks like, how they started and then continued, and how they checked their structure.

✳ **BIG IDEA.** **Shapes can be located in space and relocated by using mathematical processes.**

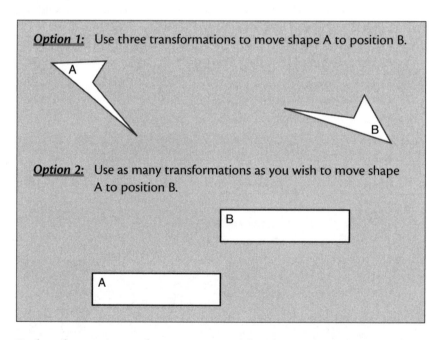

Both tasks require students to use transformations. **_Option 1_** stipulates the number of transformations to be used, which adds to the complexity of the problem. Three reflections can be used to accomplish the first task; the first moves one point on the original shape to its image, and the next two together result in the required turn:

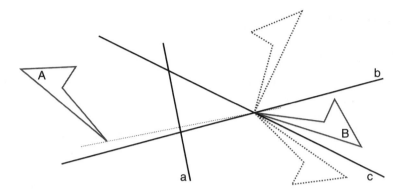

Whichever task was performed, in follow-up discussion students could be asked how they recognized which transformations to try and how they tested their predictions. Further, they could be asked to describe, using appropriate mathematical language, the transformations they used.

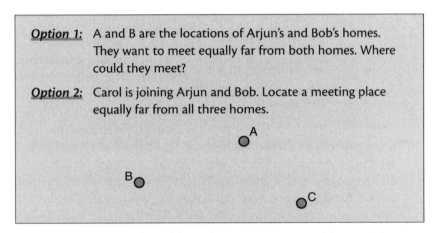

Option 1: A and B are the locations of Arjun's and Bob's homes. They want to meet equally far from both homes. Where could they meet?

Option 2: Carol is joining Arjun and Bob. Locate a meeting place equally far from all three homes.

One of the concepts in geometry that students will meet in higher grades is the concept of **locus**—identifying sets of points that meet specific conditions. These simple problems are examples of locus problems.

Option 1 asks students to locate a point equidistant from two given points, whereas *Option 2* asks students to locate a point **equidistant** from three given points. Until students explore the problem, they may not realize that there are many solutions to *Option 1* but only one solution to *Option 2*.

The solutions to *Option 1* are all the points on a line perpendicular to line segment AB that meets it at the center of the segment. The solution to *Option 2* is the single point that is the center of the one circle that passes through points A, B, and C.

Whichever task students complete, the teacher could ask:

- *Is there more than one possible meeting spot?*
- *How do you know?*
- *How did you find the meeting spot?*

SUMMING UP

MY OWN QUESTIONS AND TASKS

Lesson Goal: Grade Level: _____

Standard(s) Addressed:

Underlying Big Idea(s):

Open Question(s):

Parallel Tasks:
Option 1:

Option 2:

Principles to Keep in Mind:
- All open questions must allow for correct responses at a variety of levels.
- Parallel tasks need to be created with variations that allow struggling students to be successful and proficient students to be challenged.
- Questions and tasks should be constructed in such a way that will allow all students to participate together in follow-up discussions.

The four big ideas that underpin work in Geometry were explored in this chapter through nearly 50 examples of open questions and parallel tasks, as well as variations of them. The instructional examples provided were designed to support differentiated instruction for students at different developmental levels, targeting three separate grade bands: pre-K–grade 2, grades 3–5, and grades 6–8.

Sometimes a student struggling with Number and Operations performs well in Geometry or vice versa. Rather than assuming that certain students need scaffolding and others do not based on their performance in Number and Operations, teachers can use differentiated questions and tasks, as suggested, to allow students to work at an appropriate level.

The examples presented in this chapter only scratch the surface of possible questions and tasks that can be used to differentiate instruction in Geometry. Other questions and tasks might be created by building shapes by putting together different numbers of shapes than suggested in the examples, by applying different transformations than in the example situations, or by requiring different construction conditions for shapes than those in the examples. A form such as the one shown here can serve as a convenient template for creating your own open questions and parallel tasks. The Appendix includes a full-size blank form and tips for using it to design customized teaching materials.

Measurement

DIFFERENTIATED LEARNING activities in measurement are derived from applying the NCTM process standards of problem solving, reasoning and proof, communicating, connecting, and representing to content goals of the NCTM Measurement Standard, including

- understanding measurable attributes of objects and the units, systems, and processes of measurement
- applying appropriate techniques, tools, and formulas to determine measurements

TOPICS

Although students typically learn length before other measurements, there should be the same sort of development in teaching almost any type of measurement: moving from comparisons based on a particular attribute, to the use of nonstandard and then standard units, and then often to the use of formulas as a shortcut for measurement calculation. The NCTM Curriculum Focal Points (NCTM, 2006), which suggest what mathematical content should be the focus of each grade level, are useful for seeing which measurement attributes are usually considered at which grade levels. These guidelines can help make a teacher aware of where students' learning is situated in relation to typical earlier and later learning in measurement.

Prekindergarten–Grade 2

Within this grade band, students focus on measurement of length, initially comparing lengths either directly or indirectly and later using nonstandard and nonmetric and metric **units** to measure length. When using units, they begin to recognize that a larger unit results in a smaller numerical value for the measurement. There is also some comparison and ordering of objects in terms of weight.

Grades 3–5

Within this grade band, students begin to use fractions of units in measuring length and measure perimeter. They learn what area means and begin to measure area. They begin to develop and use formulas for areas of simple shapes, such as rectangles, and learn that 1 square inch is a useful unit for measuring and describing areas.

Students also learn about volume and **surface area** of 3-D objects and learn to calculate and estimate volumes and surface areas of simple objects.

Grades 6–8

Within this grade band, students continue to solve problems involving length, area, and volume. They develop and use formulas for surface areas and volumes of prisms and cylinders and for areas and **circumferences** of circles.

Students in this grade band begin to recognize the relationships between measurements of similar shapes. They also use the Pythagorean theorem to simplify the measurement of lengths of the sides of right triangles.

THE BIG IDEAS FOR MEASUREMENT

Coherent curricula in measurement that meet NCTM content and process standards (NCTM, 2000) and support differentiated instruction can be structured around the following big ideas:

- A measurement is a comparison of the size of one object with the size of another.
- The same object can be described by using different measurements.
- The numerical value attached to a measurement is relative to the measurement unit.
- Units of different sizes and tools of different types allow us to measure with different levels of precision.
- The use of **standard measurement** units simplifies communication about the size of objects.
- Knowledge of the size of **benchmarks** assists in measuring.
- Measurement formulas allow us to rely on measurements that are simpler to access to calculate measurements that are more complicated to access.

The tasks set out and the questions asked while teaching measurement should be developed to reinforce these ideas. The following sections present numerous examples of application of open questions and parallel tasks in development of differentiated instruction in these big ideas across three grade bands.

OPEN QUESTIONS FOR PREKINDERGARTEN–GRADE 2

> **OPEN QUESTIONS** are broad-based questions that invite meaningful responses from students at many developmental levels.

✳ **BIG IDEA.** **A measurement is a comparison of the size of one object with the size of another.**

> What can you find in the classroom that is about as long as your arm?

This question allows the teacher to see what students know about length comparisons. Watching how students compare their arm with other objects shows whether they understand the need for a **baseline** to compare length. Use of the term *about* provides enough latitude that even students who struggle can answer successfully.

> **TEACHING TIP.** When students are working with length comparisons, it is useful to have spools of string or colored yarn available for them to work with.

Variations. This task can be adapted by having students look for objects that are either *shorter than* or *longer than* another item instead of *about the same length*. The item that is used as the basis for comparison can also be varied.

> One length looks shorter than another, but it really isn't. How is that possible?

An important part of comparing lengths is recognizing that both items must be "straightened out" before they can be compared. For example, a very long coiled rope can look shorter than one that is straight but actually shorter.

Instead of setting out a coiled rope and a straight rope for students to compare, an open question such as the one proposed allows students to imagine all sorts of possible scenarios, providing access to a broader range of students. For example, a student who might have difficulty with the idea of a coiled rope being shorter might be able to imagine an object that is bent around a corner with only some of it visible.

> You want to describe to someone how long your pencil is. You can't show the person the pencil, but have to use words. How could you help the person understand its size?

This question is designed to help students see that the way a measurement is described is invariably in relation to another measurement. If, for example, a student says how many inches long the pencil is, he or she is comparing it to 1 inch. If it is described as being about as long as a hand, the comparison is being made to a hand. In fact, the only way measurements can be described meaningfully is through comparison.

The question is open in that it is completely up to the student to describe the object.

To stress the concept that a comparison is being made, the teacher could explicitly ask: *What measurement comparison are you making when you say that?*

✹ **BIG IDEA.** The same object can be described by using different measurements.

> You want to measure the carpet you will need to fit under your hamster's cage. What do you need to measure and how would you measure it?

The context suggested is one to which most students can easily relate, and it allows for a variety of responses. Whereas some students might think about the amount of space (or area) the cage occupies, others might consider length measures, for example, how long or how wide the cage is.

Students have flexibility because they can imagine the cage to be any size or shape they want and can envision a carpet just fitting the cage or being much longer or wider than the cage. No matter what decision they make, they will still be considering that the cage's size can be described by using a measurement.

> Two shapes are the same size. What could they be?

Rather than suggesting that two toys or two tables or two shoes are the same size, the question asks about two shapes. In this way, students are likely to feel less constrained about what the items look like and what aspect of their sizes might be the same. Students might imagine two shapes that are the same length, but different in width; the same area, but differently shaped; or the same volume, but different in height, for example.

> **TEACHING TIP.** Although many open questions are framed in the abstract—for example, saying that "two shapes" (rather than specifying particular objects) are the same size—it is important to provide actual objects to which students can refer.

> An object is big in one way but small in another way. What might the object look like?

This task allows students to consider either 2-D shapes or 3-D objects and offers great flexibility in the types of comparisons that can be made.

Some students might take the condition to mean that the object is big when compared with another object but small when compared with a third object. For example, a pencil is big when compared with a gnat, but small in relation to the length of a hallway. Other students might consider different attributes of the same object. For example, a very tall, very thin cylinder could be considered big in terms of height but small in terms of width. Still other students might consider two different types of measurement. For example, an object could be very heavy (i.e., big in weight) but still very small in width or length or height.

The class discussion should provide the opportunity for the teacher to explicitly make the point that when the size of an object is described, it is important to be very clear about what attribute of the object is being considered.

✳ **BIG IDEA.** The numerical value attached to a measurement is relative to the measurement unit.

> How many baby steps are there in a giant step?

In the earlier grades, students begin to understand how units can be used to describe measurements by working with **nonstandard units**. Using a unit of a baby step to measure a giant step is meaningful to students. Because of the diversity of possible correct answers, this focus on nonstandard units provides the opportunity for students to later value the usefulness of standard units.

In the short run, this question is easy for any student to act out. It also provides some flexibility, because students are free to interpret baby steps and giant steps in any way they wish.

Variations. Variations of the question can be created by using other familiar measures that come in two sizes, for example, *How many baby cups full of water do you need to fill a large glass of water?*

✳ **BIG IDEA.** **Knowledge of the size of benchmarks assists in measuring.**

> Describe three things that weigh less than a shoe. Tell how you know
> they weigh less than a shoe.

Asking for items that weigh less than a shoe provides a great deal of latitude for students; they only need to think of very light items. Some students will think of food items, others classroom supplies, others pieces of clothing, and others items such as toys.

In asking how students know the item weighs less than a shoe, it does not matter what items students chose. They will show their understanding or lack of understanding about how to use a balance or scale to compare weights.

Variations. Variations of the question can be created by suggesting an item other than a shoe or by allowing students to find heavier items rather than lighter ones.

> What lengths could you make
> by combining these strips
> in different ways?
>
>   2"
>   3"
>   2"
>   4"
>   1"

This open question is an obvious link to the Number and Operations strand. By keeping the question open and not specifying how many combinations are required or how many lengths must be combined, students are allowed to answer according to their level of understanding. Some students might choose a very simple combination of two lengths (e.g., 2 + 3 or 2 + 2), whereas others might look for many more possible combinations. Some students might even think about combining strips in zigzag paths and measuring the vertical space they occupy instead of their total length, for example:

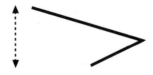

OPEN QUESTIONS FOR GRADES 3–5

✳ **BIG IDEA.** Any measurement is a comparison of the size of one object with the size of another.

> You draw a letter E, and it has an area of about 4 square units. Draw another letter and estimate its area.

There are many possible responses students might come up with. A student might choose a much smaller or much larger letter or might try to draw something very close in size to the E. A student might even turn the E sideways and call it a W.

By allowing the student to choose the second letter and its size, there is much less risk to the student of being wrong. Yet the student still applies his or her understanding of what area means.

TEACHING TIP. Differentiated instruction depends on providing some student choice. In this particular task, the student is allowed to choose the letter with which to compare the E.

> Look at a page of a newspaper that has both advertising and news. Which area is greater—the area for the ads or the area for the news?

Area becomes a more important topic in this grade band. As with other measurements, it makes sense to begin the study of area by directly or indirectly comparing measurements rather than using units. In this open question, students have a choice of comparing the measurements in whatever way they wish. Some might cut up the page and overlap the various parts. Others might, independently, come up with the idea of using some sort of unit of measurement.

Follow-up conversation among all the students will allow each one to learn how others approached the problem. In this way, all students are likely to expand their understanding of the concept of area.

Variations. Rather than comparing parts of a newspaper, students can be asked to compare other areas that might be interesting to them, for example, parts of a board game that are of different colors.

✳ **BIG IDEA.** The same object can be described by using different measurements.

How can you measure a pumpkin?

A student can think of many different ways to measure a pumpkin: width, height, weight, or circumference. Leaving the question open allows each student to choose a measurement on which to concentrate.

Class discussion about the various types of measurements students used will allow them to recognize the fact that there is usually more than one way to measure an object.

Variations. Rather than a pumpkin, another interesting object can be selected. For example, students can be asked to measure a bag of popcorn, a car, or a container.

TEACHING TIP. Items to be measured can be chosen based on themes being pursued in other subject areas of instruction, special occasions, or holidays.

✳ **BIG IDEA.** The numerical value attached to a measurement is relative to the measurement unit.

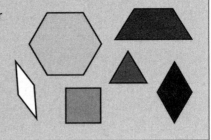

Choose a number of **pattern blocks** of the same color.

- How can you put them together around a single point so that there is no empty space?

- What does that tell you about how big the angles are that you put together?

Many times the teaching of angles is introduced with standard units. There is value, however, in thinking about nonstandard units to get across the concept of angle measurement without the additional complication of trying to convey what a degree is. Putting together many copies of the same angle, as is suggested in this task, helps students think about an angle as a nonstandard unit.

Allowing students to choose which shape they use gives them the chance to pick a simple angle (such as a right angle in the square) or a more complex angle (e.g., one of the trapezoid angles). A connection to degrees can also be made. Students will notice that 3 hexagons can be put together, so the angle measure is $360 \div 3 = 120°$; 6 triangles can be put together, for an angle measure of $360 \div 6 = 60°$; 4 squares can be put together, so the angle measure is $360 \div 4 = 90°$; either 3 or 6 **rhombuses** can be put together, depending on which angles are used; and 12 of the thin kites can be put together if the small angles are used, for an angle measure of $360 \div 12 = 30°$.

✹ BIG IDEA. **Units of different sizes and tools of different types allow us to measure with different levels of precision.**

> Your grandfather is building you a miniature car out of wood. What units should he use to measure the parts of the car? Why?

Students might have differing opinions on what a miniature is—is it very tiny, such as one that would fit in a pencil case, or is it the size a baby could sit in? This decision on scale could affect the student's choice of unit.

Many students relate the size of a unit only to the size of what is being measured; they would automatically use a small unit for a small object and a large unit for a large object. Although doing this makes sense, the teacher should help students understand that the precision needed in the measurement is also important. Even when measuring something that is longer than a ruler, one might decide to measure in millimeters because of a need to be precise. Students should consider whether the construction project would or would not require this level of precision.

The flexibility in what the miniature could look like and how precise the measurements would need to be makes the question open to a broad range of students.

Variations. The task can be varied by asking students to consider what tool or unit might be used to measure a long distance, for example, the distance from one side of the school to the other side.

✹ BIG IDEA. **Knowledge of the size of benchmarks assists in measuring.**

> A container holds about 4 gallons. Describe its size in other ways.

Some students will answer the question by relating the 4 gallons to familiar, known containers. For example, if a student knows that a kitchen pot at home holds 1 gallon, he or she can imagine another pot that holds 4 times as much. Other students will answer the question by suggesting width, depth, or height measurements that are reasonable for a container of this size.

Discussing the various approaches to the problem will help all students understand how knowing the size of one item can assist in determining the size of other items.

> Choose a pattern block. Create a shape made of 20 of your blocks.
>
> Now choose another block. How many blocks of this type will you need to make a shape that takes up the same space?

This question is designed to allow students to create their own measurement benchmarks. Students can choose to make their first shape with small blocks so that they will need fewer of the alternate blocks, making it easier to estimate. Or they might choose the challenge of using larger blocks at first.

One solution might be, for example, that the space occupied by 20 yellow blocks is the same as the space occupied by 40 red blocks. Another is that the space occupied by 20 triangles is the same as the space occupied by 10 rhombuses.

TEACHING TIP. Pattern blocks are a useful tool for teaching concepts in number (particularly about fractions and ratio), geometry, and measurement.

> Why might it be useful to measure the length and width of a room by counting how many steps you need to get from one wall to the next?

Students begin to believe, often as a result of instruction, that once they know how to measure in inches, feet, yards, meters, or centimeters, there is no longer any point in measuring with nonstandard units. This question offers them the opportunity to confront that belief without actually telling them to do so. For example, a student might recognize that if his or her parents were measuring for carpet and only needed to estimate the cost, it would be quicker and easier to walk the distance than to take the time to measure exactly.

OPEN QUESTIONS FOR GRADES 6–8

✷ **BIG IDEA.** The same object can be described by using different measurements.

> A shape has an area of 200 square inches. What could its length and width be?

By not specifying what the shape is, this question is much more open than it would otherwise be. Some students will decide to use a square. They may struggle a bit because 200 is not a **square number**, but they may recognize that the square's length and width would be about 14 inches. Other students will choose a simpler strategy, using a rectangle of length 20 and width 10. Still other students will choose a shape with more sides, particularly if grid paper is provided that would allow them to experiment. Students should be encouraged to determine more than one answer so that they use a variety of formulas or alternate methods to create shapes.

Variations. The question can be varied by changing the area required or by adding stipulations to the type of allowable shapes. For example, it could be specified that the shape must have four or more sides or that at least two of its sides must be congruent. Another way to vary the question is to give a volume measurement rather than an area measurement.

TEACHING TIP. By providing the area of a shape and not specifying additional information about the shape, students are free to access area formulas that are familiar to them. During the follow-up discussion, they may be exposed to formulas with which they are less comfortable.

You create the **net** of a 3-D figure and calculate its area. Then you fold the shape into the 3-D figure and calculate its volume.

For example, for this net, the area is 16 square units and the volume is 4 cubic units.

- Which is usually greater—the number for the area or the number for the volume? (This time the number for the area is greater.)

- Why does that make sense?

This open question requires students to make a conjecture. Some will try perhaps one other shape, others will recognize the value of trying many more shapes, and a few students may apply algebraic thinking to solve the problem. To think about why their answer makes sense, students need to either visualize a variety of situations or use formulas to help them understand.

Students might learn that if one of the dimensions of the prism is 1 unit, the surface area is always a greater number. Otherwise, either value can be greater. For example, for a $10 \times 10 \times 20$ prism, the volume is 2,000 cubic units, but the surface area is only 1,000 square units. For a $3 \times 4 \times 5$ prism, the volume is 60 cubic units, but the surface area is 94 square units.

The areas of two shapes are almost the same, but the perimeters are very different. What might the shapes be?

It is not uncommon to have students compare the areas of two rectangles with the same perimeter or the perimeter of two rectangles with the same area. By not specifying the shapes in the question and not requiring that the shapes be the same, students have much more freedom to work with shapes of their own choice. Additional flexibility is provided by suggesting that the areas are almost the same, rather than exactly the same.

As students create the required shapes, they will likely get a great deal of practice with calculating perimeters and areas. One solution is shown here:

✹ **BIG IDEA.** **Knowledge of the size of benchmarks assists in measuring.**

About how many people do you think can fit in our school?

This question is a very open one. Students can choose how close together the people can be, how to estimate the size of the school, whether there is furniture in the school that must be considered, and whether the people are big or small. With this amount of freedom, the discussion of the problem becomes much more interesting.

Students are likely to use a small area of their own classroom as a benchmark, and then must determine how to use this information to consider the larger space of the whole school. Whereas some students will be systematic, counting the number of classrooms while also considering other types of rooms such as the gym and hallways, other students will think less comprehensively. Because only an estimate is required, the range of possible answers is great.

An object has a volume of 1 cubic yard. What might it be?

By encouraging students to become aware of some familiar volume measurements, the teacher will make it easier for them to estimate the volumes of other objects. For example, if a student knows that 1 cubic yard might be the volume of a box that holds a large TV set, they will be in a much better position to estimate, for example, the volume of a room.

The question is open in that students are free to think of any object they wish. Some might think of a pile of leaves or the space inside a car or truck; or others might use yardsticks and measure a space in the classroom.

✴ BIG IDEA. **Measurement formulas allow us to rely on measurements that are simpler to access to calculate measurements that are more complicated to access.**

> A shape has the same area as this one. What could its dimensions be?

Students are free to calculate the area of the triangle in a variety of ways. Some might measure the base and height and use the formula for area; others might superimpose the triangle on a grid and calculate the area that way.

Some students will realize that they do not need to measure at all. They could simply trace and cut the triangle into two pieces and rearrange the pieces to create a new shape.

Other students will think more algebraically and recognize that they do not need to measure. They could consider the formula A = $bh \div 2$ and realize they could simply double the base and halve the height to end up with the same area.

In discussing the use of the formula, the teacher could emphasize how simple it is to measure the length of the base or height as opposed to needing a grid to determine the triangle's area. This emphasis helps reinforce for students the power of measurement formulas.

A circle is placed inside a square so that it touches the square on all four sides. About how much of the square does the circle take up?

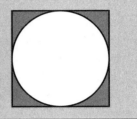

This question is open in that it allows a variety of approaches. Some students might use the formulas for the area of a circle and a square. They will realize they are comparing πr^2 to $(2r)^2$, so the ratio is $\frac{\pi}{4}$, or about $\frac{3}{4}$. Other students might cut out a figure like the one shown, transfer it to a grid, and count grid squares. Still other students might estimate that the pieces at the corner look like about one fourth of the total area of the square, and therefore estimate the circle to be $\frac{3}{4}$ of the total area.

Students might record the ratio as a fraction, a decimal, or a percentage. They might try many circles and squares or might assume that what is true in one case is more generally true.

TEACHING TIP. Questions that have students compare measurements are often more valuable for students than questions involving only one measurement. They get double the practice, and the notion of measurement as a comparison is reinforced.

You start with a parallelogram. You increase its height by the same amount as you decrease its base length.

How does the area change?

To respond to this question, some students will use a formula. For example, a student who realizes that $A = bh$ might try various number combinations for b and h to see what happens:

b	4	3	2	1
h	3	4	5	6
A	12	12	10	6

In the table above, students can see that the area seems to go down as b and h become more different, but they are likely to try other combinations.

Other students will simply draw parallelograms on a grid and determine what happens to the area.

Some students will increase the height and decrease the base by a fixed amount, whether 1 unit or some other amount, whereas other students will try different amounts of increase and decrease. Some students will start with parallelograms with base and height dimensions that are close to one another, whereas others will start with a base much greater than the height or a height much greater than the base. Regardless of how many examples they try or what method they use, all students will practice the measurement of area and all will have the opportunity to learn that something different can happen depending on the value of $b - h$.

PARALLEL TASKS FOR PREKINDERGARTEN–GRADE 2

PARALLEL TASKS are sets of two or more related tasks that explore the same big idea but are designed to suit the needs of students at different developmental levels. The tasks are similar enough in context that all students can participate fully in a single follow-up discussion.

✳ **BIG IDEA.** A measurement is a comparison of the size of one object with the size of another.

> Decide which is longer.
>
> **Option 1:** The distance from your shoulder to your wrist or the distance around your head
>
> **Option 2:** The distance from your elbow to your wrist or the length of your foot

These tasks differ in that students can literally move their foot up to their arm to compare lengths for **Option 2** but need to use an indirect method to compare measurements (e.g., using a string) for **Option 1**.

Whichever option was chosen, students could be asked:

- *Can you just look to decide which is longer?*
- *What do you have to do to check?*
- *What other length comparisons would be easy to make?*
- *Which would be trickier?*

TEACHING TIP. When numbered options are offered, sometimes the "simpler" option should be presented as Option 1 and other times as Option 2. The unpredictability will ensure that students consider both possibilities when they choose their tasks.

✹ BIG IDEA. **The same object can be described by using different measurements.**

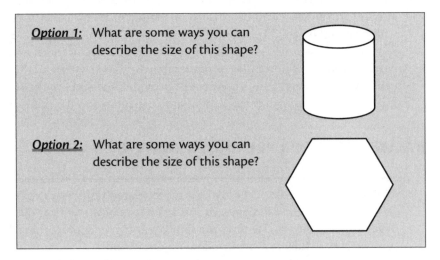

Option 1: What are some ways you can describe the size of this shape?

Option 2: What are some ways you can describe the size of this shape?

Some students will be more comfortable describing the size of a 2-D shape, whereas others will be more comfortable with a 3-D shape. In the case of the hexagon, students might consider the width, the height, the lengths of diagonals, the space it occupies, or the length of each side. In the case of the cylinder, students might consider the height, the width, the distance around, or maybe even the thickness of the shape if it is seen as hollow.

Whichever option students chose, it would be meaningful to ask:

- *Would you consider the size large or small or neither?*
- *How could you compare the size of this shape with other shapes you know?*
- *How could you decide if your shape is bigger or smaller than another shape?*

Variations. This question can be reused by specifying other pairs of shapes: one 3-D and the other 2-D, or perhaps two 2-D shapes with one more complex than the other.

Option 1: Describe an object that is small but might weigh a lot.

Option 2: Describe an object that is very tall, but is not very wide.

Both of these tasks help students to see that an individual object can be measured in more than one way. ***Option 2*** involves only length measures, whereas ***Option 1*** one involves both weight and another measure (e.g., length).

Providing a choice allows the student who might find it difficult to visualize a heavy or light object to focus only on length.

In either case, students can build on their everyday life experience to help them answer the question. Potential objects for ***Option 1*** might be a heavy metal ball or a small rock. Objects for ***Option 2*** might include a length of string or ribbon or the crack between the two parts of a folding door.

Parallel Tasks for Pre-K–Grade 2

No matter which option was selected, a student could be asked:

* *What object did you choose? How does it fit the rule for its size?*
* *Why is it probably not a good idea to tell someone "I'm thinking of something small" and assume they will know what you mean?*

TEACHING TIP. Questions that build on students' everyday lives help them see the value of understanding mathematical ideas.

✺ **BIG IDEA.** **The numerical value attached to a measurement is relative to the measurement unit.**

> ***Option 1:*** Find five items in the room that are between 10 cm and 25 cm long.
>
> ***Option 2:*** Find five items in the room that are shorter than your pencil.

Some students will be comfortable with standard units of measure, whereas others will still be at the stage where comparison, rather than the use of units, is more meaningful. The choice offered in the two options allows each group of students to attack a problem at an appropriate level.

Rulers can be provided for ***Option 1***, although students might first estimate without using a ruler and then check their estimates. Similarly, students might take a pencil with them as they attempt ***Option 2*** or they might, instead, estimate and then check once they have chosen their items.

Questions that would be appropriate for both groups include:

* *What items did you find?*
* *How do you know that your items meet the required rule?*

✺ **BIG IDEA.** **Knowledge of the size of benchmarks assists in measuring.**

> A table is 5 pencils long.
>
> ***Option 1:*** How many paper clips long would it be?
>
> ***Option 2:*** How many inches long would it be?

A pencil is a familiar item for students, so it serves as a meaningful benchmark. Some students will easily compare that benchmark with another nonstandard unit, such as a paper clip, whereas other students will be comfortable comparing that length with a standard unit such as inches. Thus, the choice provided in the two options serves the needs of a broad range of students.

Questions that could be asked of students who chose either option include:

- *Is the table a long one or a short one?*
- *How do you know? How did you decide how long the table was using a different unit?*
- *Was the number greater than or less than 5? Why?*

Use a **pan balance** to make a modeling clay ball with the same mass as a toy car.

Option 1:　Now make a clay ball with the mass of four toy cars.

Option 2:　Now make a small clay ball so that four of them, altogether, have the mass of the toy car.

In both instances, students use a referent or benchmark, the original clay ball, to create other masses. In **_Option 1_**, students must multiply their mass. In **_Option 2_**, students must divide their mass. Some students might find one of the tasks easier than the other.

All students could be asked:

- *Is your new ball heavier or lighter than the original one? Why?*
- *Did your ball have to be perfectly round to be the right weight or could it have been slightly flattened?*
- *How did you create the second clay ball?*

PARALLEL TASKS FOR GRADES 3–5

✴ **BIG IDEA.**　**Any measurement is a comparison of the size of one object with the size of another.**

Use this grid of dots.

Option 1:　Make as many shapes as you can on the grid with an area of 12. The corners of the shapes must be dots on the grid.

Option 2:　Make as many rectangles as you can on the grid with an area of 12. The corners of the rectangles must be dots on the grid.

Both of these tasks require students to consider area. Both allow for some latitude by asking students to construct as many shapes as they can, rather than a fixed number. In this way, a student who can find only one or two shapes will still feel successful.

Some students may decide not to use congruent shapes. For example, if they place a 4 by 3 rectangle in one position, they are not likely to create another 4 by 3 rectangle in another position. Encourage them to do this to help build their spatial skills.

Some students might find the first option easier because they have latitude in the shape they can form. They can simply connect 12 squares in some fashion to make a shape. However, it will be hard for them to be organized to produce as many shapes as possible. Other students will recognize that the rectangles required in **_Option 2_** must be 4 by 3 rectangles and will spend their energy on ensuring that they consider all the possible placements of the rectangles.

Students could be asked:

* _How did you find your first shape?_
* _How did you use that first shape to help you get other shapes?_
* _How do you know that your shape has area 12?_
* _Are there other shapes with area 12? How do you know?_

✹ BIG IDEA. **The same object can be described by using different measurements.**

> **_Option 1:_** A rectangle has sides that are whole numbers of inches long. The perimeter is 44 inches. Draw five possible shapes.
>
> **_Option 2:_** A polygon has a perimeter of 44 inches. Draw five possible shapes.

Clearly, **_Option 2_** provides more choice than **_Option 1_**. Some students will appreciate the choice, recognizing that they need only take a piece of string 44 inches long and reshape it into various polygons. Other students will feel more comfortable working with rectangles only. Once they realize that the length and width are each represented twice in a perimeter and that therefore the length plus the width must be 22, students can simply use arithmetic skills to build their rectangles.

All students could be asked:

* _What does it mean to know that the perimeter of a shape is 44 units?_
* _How did you select your first shape?_
* _How do you know that your perimeter is 44 inches?_
* _How did you use your first shape to help you find other possible shapes?_

For each puzzle, the set of all the missing numbers is listed below the puzzle. It is your job to figure out which number goes in which spot. Complete the puzzle.

Option 1: A cereal box is _____ cm high, _____ mm deep, and _____ cm wide. It holds about _____ pieces of cereal.

19   30   73   950

CEREAL

Option 2: A jet can fly _____ miles each hour. This is _____ times the distance a car on a highway travels in the same amount of time. It can fly as high as about _____ miles in the air. This is about the height of _____ Empire State Buildings.

7   9   25   550

Both of these tasks require students to use **proportional thinking**. **Option 1** might be easier for many students because it is a more concrete task; they can more easily imagine the size of the cereal box from their own life experience and can use the visual to help as well. On the other hand, **Option 1** involves measurement conversions, in this case centimeters to millimeters. **Option 2** might be more interesting to students, however, because it provides them with new information.

The solution to **Option 1** is (in order) 30, 73, 19, and 950; the solution to **Option 2** is 550, 9, 7, and 25.

Both groups could be asked questions such as:

- *What number were you sure of first? Why?*
- *Which number was hardest for you to get? Why?*
- *How could you be sure that your numbers made sense when you placed them in the blanks?*

Variations. It is not difficult to find other objects with multiple measurements that will be familiar to students and to make up other fill-in-the-blank questions involving those measurements.

> **_Option 1:_** About how much is a line of pennies that is 1 mile long worth?
>
> **_Option 2:_** About how much is a line of pennies that is 1 yard long worth?

Students can line up real pennies against a ruler to get a sense of how much a line of pennies that is 1 foot long is worth. Then they can use calculations to help them respond to one of the given tasks. Clearly, **_Option 2_** is easier because students need only to multiply a relatively small number by 3, rather than by 5,280.

Whichever option was selected, students could respond to questions such as these:

- *About how wide is one penny?*
- *Is that information useful? How?*
- *About how much would a line of 1,000 pennies be worth?*
- *How do you know?*
- *How could you check your results without actually lining up all the pennies?*

✸ BIG IDEA. Units of different sizes and tools of different types allow us to measure with different levels of precision.

> **_Option 1:_** Describe something you would probably measure in miles and something else you would probably measure in yards.
>
> **_Option 2:_** Describe something you would probably measure in inches and something else you would probably measure in centimeters.

Each of the tasks requires students to make a judgment about what unit is more appropriate in a situation. In **_Option 1_**, the decision is more obvious because the difference between miles and yards is so great. In **_Option 2_**, the decision is more subtle. Some students will respond in **_Option 2_** by referring to their measuring tool and the units that are available on it, whereas others will consider the size of the objects to be measured. Recognizing that the centimeter is smaller than the inch, they will likely choose the centimeter for the smaller item.

Both groups of students could be asked:

- *Both of your units are length units. Could you use either unit to measure any length?*
- *What influences your decision about which unit to use?*

Estimate the size of each footprint.

Option 1: Count the number of squares that the footprint fully covers. Count the number of squares that the footprint partially or fully covers. Select the number halfway between the two.

Option 2: Count the number of squares that are more than half covered by the footprint.

Option 3: Cover the footprint with aquarium gravel, one layer thick. Rearrange the gravel into a rectangle to estimate the area of the footprint.

All three options require students to consider what it means to estimate area. **_Option 3_** is more labor intensive but is probably the most comfortable for students because it relates the area of something irregular to something with which they are much more comfortable, a rectangle. **_Options 1_** and **_2_** merely require counting. **_Option 2_** demands that students estimate whether each square is more than half covered or not; many students will find this repeated decision making uncomfortable. **_Option 1_** is somewhat more straightforward, but why a number halfway between the two values is chosen may be somewhat of a mystery.

All three groups of students could be asked:

- *Is it easy to tell which of the footprints is larger just by looking at the grid?*
- *Did your estimates help you decide which was larger?*
- *What was your estimate for each?*
- *Why did those values seem reasonable to you?*

TEACHING TIP. It is important that students get a lot of experience in measuring irregular shapes, and not just regular shapes. Real life is full of both.

✳ **BIG IDEA.** The use of standard measurement units simplifies communication about the size of objects.

> ***Option 1:*** An object has an area of 12 square inches. What might it be?
>
> ***Option 2:*** An object has an area of 12 pattern block triangles. What might it be?

The difference between the two tasks relates to the student's comfort with standard area measurements. Some students might choose ***Option 2*** because it is easier for them to visualize the size of a pattern block triangle than a square inch.

Possible objects 12 square inches in size are a cordless phone laying flat on a table or the base of a large cup. Possible objects the size of 12 pattern block triangles include a flip cell phone or a roll of tape laying on its side.

For either task, the teacher could ask:

• *How much space would one unit use up? How do you know?*
• *How did you solve the problem?*

PARALLEL TASKS FOR GRADES 6–8

✳ **BIG IDEA.** The same object can be described by using different measurements.

> ***Option 1:*** One circle has a greater area than another. Does its circumference also have to be greater?
>
> ***Option 2:*** One rectangle has a greater area than another. Does its perimeter also have to be greater?

Some students who are new to the formulas involving circumference and area of a circle are likely to be attracted to ***Option 2***, using rectangles. What is interesting, however, is that the answer to ***Option 1*** is clearer than the answer to ***Option 2***. If one circle has a greater area than another, it has a greater radius and therefore a greater circumference. On the other hand, a rectangle with a greater area can sometimes have a greater perimeter and sometimes not. For example, a 6 by 10 rectangle has more area but less perimeter than a 1 by 30 rectangle, but a 4 by 5 rectangle has more area and more perimeter than a 1 by 2 rectangle.

Students completing either task could be asked:

• *How many combinations did you try?*
• *Can you be sure of the result with that many trials?*
• *How could a picture help someone understand your thinking?*
• *How might you have predicted what you found out?*

Each prism has whole number unit side lengths.

Option 1: A prism has a surface area of 126 square units and a volume of 90 cubic units. What could the dimensions be?

Option 2: A prism has a volume of 48 cubic units and a width of 4 units. What could the dimensions be?

Option 2 will be easier for many students than **Option 1** because one of the three dimensions is provided. Students choosing **Option 2** will realize that the product of the length and height must be 12 square units and can simply list some possibilities. The challenge of **Option 1** requires students to simultaneously apply formulas for both surface area and volume.

Regardless of which option the students chose, it would be meaningful to ask:

- *What is the relationship between the dimensions and the volume?*
- *What is the relationship between the dimensions and the surface area?*
- *Can you be sure that there are no other possible dimensions?*

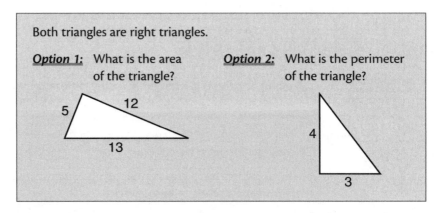

Both triangles are right triangles.

Option 1: What is the area of the triangle?

Option 2: What is the perimeter of the triangle?

Option 1 is set up with all of the required information provided, as long as students recognize that orientation is irrelevant in describing the base and height of a triangle. Here the base could be viewed as the side with length 5 and the height as length 12; the area, therefore, is 30 square units. **Option 2** requires application of the Pythagorean theorem.

Whichever task was completed, students could be asked:

- *You calculated one measure of the shape. How did you do the calculation?*
- *Why does that method make sense?*
- *What if you had been asked to calculate a different measure?*
- *Would it have been equally easy?*

✳ **BIG IDEA.** **Knowledge of the size of benchmarks assists in measuring.**

> A prism has a volume of about 200 cm³.
>
> ***Option 1:*** Describe the dimensions of a different prism with about the same volume.
>
> ***Option 2:*** Describe the dimensions of a cylinder with about the same volume.

Both options require that students have an understanding of the formula for the volume of a prism, but ***Option 2*** also requires that students have either a formal or informal understanding of the formula for the volume of a cylinder. A student who chooses ***Option 1*** could look for groups of three numbers that multiply to produce 200, for example, 10 by 20 by 1, or 4 by 10 by 5. A student who chooses ***Option 2*** might recognize that a prism with a base area of 20 and a height of 10 cm would have the desired volume. They might then choose a cylinder with a slightly larger base and similar height, figuring that the cylinder needs to be a bit wider to make up for the fact that a circle takes up less space than a square of the same width.

Both groups of students could be asked:

- *Can you be sure of the dimensions of the original prism? Why not?*
- *Is there a maximum height it could be? A minimum height? Why not?*
- *Is it helpful to know volume formulas to help you solve the problem? How?*

✳ **BIG IDEA.** **Measurement formulas allow us to rely on measurements that are simpler to access to calculate measurements that are more complicated to access.**

> Circular targets often are made up of circles of alternating colors.
>
>
>
> ***Option 1:*** What could be the widths of the rings if the black ring is $\frac{1}{4}$ of the area of the whole circle?
>
> ***Option 2:*** The inner white circle, the outer white ring and the black ring all have equal areas. What are the widths of the white and black rings if the inner circle has a radius of 5"?

Both of these options require that students use the formula for the area of a circle. There are many more possible answers for ***Option 1*** than for ***Option 2*** partially, because no specific sizes are listed and partially because there are no requirements for how the two white sections relate. Because there are more potential answers, **_Option 1_** may be more accessible to students. However, some students may prefer **_Option 2_** because it is specific.

Whichever option was selected, a teacher could ask:

- *How could the black ring be thought of as the difference between two circles?*
- *How could the white ring be thought of as the difference between two circles?*
- *Could the inside white circle be half of the whole circle?*
- *What did you decide the widths of the rings were?*

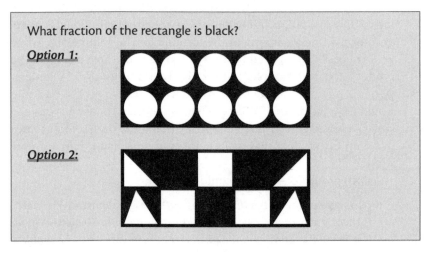

Students need to know how to calculate the area of a circle to complete **_Option 1_** (the fraction of black is actually about $\frac{1}{4}$) but only the areas of triangles and squares to complete **_Option 2_** (the fraction of black is actually $\frac{1}{2}$). In either case, students can apply their knowledge of area formulas; all that differs is which formulas they must focus on.

Whichever option was selected, a teacher could ask:

- *Is the black area more or less than half the total area? Why do you think that?*
- *What formulas did you use to help you?*
- *How much of the rectangle do you think is black?*

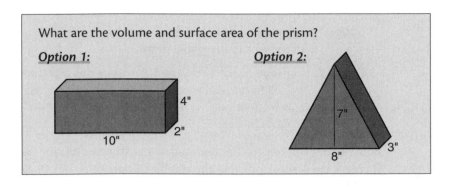

To complete **_Option 2_**, students not only need to know what volume and surface area mean and how to calculate them, but they also need to be able to use the Pythagorean theorem to determine the side length of the side rectangles. For **_Option 1_**, they need to know how to calculate surface area and volume of rectangular shapes only. The volume and surface area for **_Option 1_** are 80 cubic inches and 136 square inches; the volume and surface area for **_Option 2_** are 84 cubic inches and 128.4 square inches.

Whichever option was selected, a teacher could ask:

- *Which face has the greatest area?*
- *How do you know?*
- *Why is it hard to tell which has the greatest volume by just looking at the pictures?*
- *Does a shape that is taller always have a greater volume? A greater surface area?*
- *How did you calculate the volume? The surface area?*

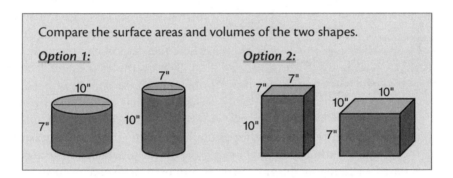

Compare the surface areas and volumes of the two shapes.

Option 1:   **_Option 2:_**

In both cases, students need to know how to calculate surface area and volume, but in one case they must be able to deal with cylinders and in the other case only with rectangular prisms. Many students will assume that the volumes are equal in **_Option 1_** because they see 7 and 10 on both shapes, but this is not the case. They may be less sure of how the surface areas compare.

In fact, for **_Option 1_**, the volumes are 175π cubic inches for the short cylinder compared to 122.5π cubic inches for the tall one, and the surface areas are 120π square inches compared to 94.5π square inches. For **_Option 2_**, the volumes are 490 cubic inches and 700 cubic inches, respectively, and the surface areas are 378 square inches and 480 square inches, respectively.

Whichever option was selected, a teacher could ask:

- *Is it possible for the volumes to be the same but the surface areas to be different?*
- *Is it possible for the surface areas to be the same but the volumes to be different?*
- *Are either the same in your pair of shapes?*
- *How did you calculate the volumes?*
- *How did you calculate the surface areas?*

SUMMING UP

MY OWN QUESTIONS AND TASKS

Lesson Goal: Grade Level: _____

Standard(s) Addressed:

Underlying Big Idea(s):

Open Question(s):

Parallel Tasks:
Option 1:

Option 2:

Principles to Keep in Mind:
- All open questions must allow for correct responses at a variety of levels.
- Parallel tasks need to be created with variations that allow struggling students to be successful and proficient students to be challenged.
- Questions and tasks should be constructed in such a way that will allow all students to participate together in follow-up discussions.

The seven big ideas that underpin work in Measurement were explored in this chapter through nearly 50 examples of open questions and parallel tasks, as well as variations of them. The instructional examples provided were designed to support differentiated instruction for students at different developmental levels, targeting three separate grade bands: pre-K–grade 2, grades 3–5, and grades 6–8.

Measurement is a strand that links Number and Operations with Geometry. Students experiencing weakness in either of those strands might struggle in Measurement. Because Measurement is a very practical part of the mathematics that is taught, it is important to maximize student success with this strand through differentiating instruction to meet students where they are developmentally.

The examples presented in this chapter only scratch the surface of possible questions and tasks that can be used to differentiate instruction in Measurement. Other questions and tasks can be created by, for example, changing the type of measurement in an example question or task, changing the object to which a measurement is compared, or changing the units that are proposed to be used. A form such as the one shown here can serve as a convenient template for creating your own open questions and parallel tasks. The Appendix includes a full-size blank form and tips for using it to design customized teaching materials.

Algebra

DIFFERENTIATED LEARNING activities in algebra are derived from applying the NCTM process standards of problem solving, reasoning and proof, communicating, connecting, and representing to content goals of the NCTM Algebra Standard, including

- understanding patterns, relations, and functions
- representing and analyzing mathematical situations and structures using algebraic symbols
- using mathematical models to represent and understand quantitative relationships
- analyzing change in various contexts

TOPICS

Before beginning the task of differentiating student learning in algebra, it is useful for teachers to have a good sense of how the topics in the strand develop over the grade bands. It is particularly important in algebra because the concept of what constitutes algebraic thinking has broadened in recent years. Work with pattern and a great deal of number work involving the use of relationships is now viewed as part of algebraic thinking.

With the notion of algebra representing general relationships and change, students use algebra in the early grades as they relate numbers additively and solve simple addition and subtraction equations, later as they relate numbers multiplicatively and solve simple multiplication and division equations, and even later as they use symbolism to describe relationships between numbers and solve more complex equations.

The NCTM Curriculum Focal Points (NCTM, 2006), which suggest what mathematical content should be the focus of each grade level, can be helpful in making teachers aware of where students' learning is situated in relation to what learning in algebra typically precedes and succeeds the work in a particular grade band.

Prekindergarten–Grade 2

Within this grade band, students identify, describe, and extend simple number and shape patterns. They use strategies involving patterns to help them learn addition and subtraction facts, and they skip count as a prelude to later work on multiplicative relationships.

Grades 3–5

Within this grade band, students continue to identify, describe, and extend number patterns, including **increasing**, **decreasing**, and **repeating patterns**. They also use patterns and relationships to help them analyze multiplication and division situations.

Grades 6–8

Within this grade band, students move into more traditional algebra. They use expressions, equations, and formulas to correspond to numerical and real-life situations. They evaluate expressions involving **variables** and use variables more regularly, recognizing that two different expressions might be equivalent. They solve simple equations and use tables of values to uncover relationships.

THE BIG IDEAS FOR ALGEBRA

Coherent curricula in algebra that meet NCTM content and process standards (NCTM, 2000) and support differentiated instruction can be structured around the following big ideas:

- A group of items form a pattern only if there is an element of repetition, or regularity, that can be described with a pattern rule.
- Any pattern, **algebraic expression**, relationship, or equation can be represented in many ways.
- Patterns are all around us in the everyday world.
- Many number, geometry, and measurement ideas are based on patterns.
- Arranging information in charts and tables can make patterns easier to see.
- Variables can be used to describe relationships.

The tasks set out and the questions asked while teaching algebra should be developed to reinforce these ideas. The following sections present numerous examples of application of open questions and parallel tasks in development of differentiated instruction in these big ideas across three grade bands.

OPEN QUESTIONS FOR PREKINDERGARTEN–GRADE 2

OPEN QUESTIONS are broad-based questions that invite meaningful responses from students at many developmental levels.

✸ **BIG IDEA.** **A group of items form a pattern only if there is an element of repetition, or regularity, that can be described with a pattern rule.**

> Take 24 counters. Use 12 of them to make a pattern and the other 12 to make a nonpattern. Tell why one is a pattern and the other is not.

This task requires students to show that they understand the concept of regularity as being critical to patterns. For example, a student might make these two sequences of counters:

Although technically the second sequence could be the beginning of a larger pattern, most students at this level will just look at what is there and decide it is not a pattern. Encouraging students to tell why the first line shows a pattern and the second does not is a valuable exercise. The question is suitable for a broad range of students because students can use as simple a pattern rule as they wish to create their pattern.

TEACHING TIP. It is easier to create an open question by asking students to create a pattern than it is to extend a given one.

> The fourth picture in a pattern consists of five squares as shown:
>
> | ? | | ? | | ? | | ⬜⬜⬜ ⬜ ⬜ | | ? |
>
> What could the first, second, third, and fifth pictures look like?

Students have a significant amount of flexibility in answering this question. They can create a repeating pattern, an increasing pattern, or even a decreasing pattern, whichever they prefer working with. Some students might show a pattern such as this:

Others might produce a pattern like this one:

Rather than making the fourth **term** in the question four squares, five squares were used. Showing four squares would have made it much more likely that most students would create the same 1, 2, 3, 4 pattern.

Variations. This task can be reused by changing the picture for the fourth term or by defining a different term, for example, the second or third one.

> A certain pattern has the number 5 in it. What could the pattern be?

This very open question allows students to make as simple or as complex a pattern as they wish. Many students will simply make a repeating pattern such as 2, 5, 2, 5, 2, 5, . . . ; others might create a pattern with a three-term **core**, for example, 2, 4, 5, 2, 4, 5, 2, 4, 5, ... Still other students will produce an increasing pattern, such as 2, 3, 4, 5, ...

No matter what pattern students create, it is important that they be able to respond to a question such as:

- *What makes this a pattern?*
- *Could you change any of the numbers and still have a pattern?*
- *How could you do that?*

✱ BIG IDEA. **Any pattern, algebraic expression, relationship, or equation can be represented in many ways.**

> Sabrina made the pattern below.
>
>
> Make a pattern that you think is like this. Tell why the patterns are alike.

Although many young children focus on the specific items that make a pattern, mathematically it is the *form* of the pattern that matters most. Once students realize that the pattern above is, essentially, the same pattern as, for example, square, circle, circle, triangle, repeated over and over, they have made an important mathematical generalization. Asking students to translate patterns to a different form is one way to access this concept.

An even more open approach is to simply ask for two patterns that are alike. Some students will create one pattern and translate it to a different form. Others will interpret *alike* in other ways. For example, they might think that two patterns are alike if they both use shapes or if they both use numbers.

TEACHING TIP. For differentiated instruction to be successful, it is essential that the teacher be flexible in accepting answers. Even if the teacher has a particular response in mind, if the student goes in a different direction—but one that makes sense—there is no need to ask the student to revisit the original question.

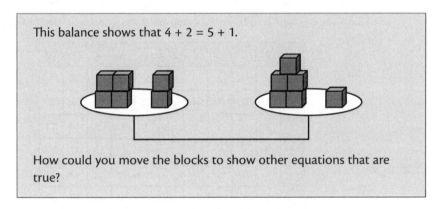

This balance shows that 4 + 2 = 5 + 1.

How could you move the blocks to show other equations that are true?

It is very important that students think about equality in a relational way, that is, an equality is a statement that two expressions are identical. This becomes the basis for many arithmetic and algebraic concepts and procedures later on. This open question allows students to begin to explore how an equation can be transformed into an **equivalent equation** by using simple operations. However, how a student does the transformation is left open to allow access to all children. Some children will rearrange the given cubes, whereas others might add to or take away cubes from both sides.

Variations. Variations of the question can be created by changing the original equation used.

✹ **BIG IDEA.** Patterns are all around us in the everyday world.

> Think of a familiar song or a poem that has a pattern. Sing it or tell it to us and tell us what the pattern is.

One of the attractions of pattern work for young children is that it reflects what they see around them. Helping students see patterns in their everyday life is important. Although a teacher could simply ask students to look for examples of patterns in the classroom, it is probably more engaging to ask for songs or poems because these can be enjoyed by the whole group.

Once students offer their poems or songs, the teacher should make sure to ask them why each song or poem has a pattern. This is necessary to emphasize the mathematical thinking that is being cultivated.

✹ **BIG IDEA.** Arranging information in charts and tables can make patterns easier to see.

> Look at the addition table. List two patterns you see in the table.
>
> | + | 0 | 1 | 2 | 3 | 4 | 5 | 6 | 7 | 8 | 9 |
> |---|---|---|---|---|---|---|---|---|---|---|
> | 0 | 0 | 1 | 2 | 3 | 4 | 5 | 6 | 7 | 8 | 9 |
> | 1 | 1 | 2 | 3 | 4 | 5 | 6 | 7 | 8 | 9 | 10 |
> | 2 | 2 | 3 | 4 | 5 | 6 | 7 | 8 | 9 | 10 | 11 |
> | 3 | 3 | 4 | 5 | 6 | 7 | 8 | 9 | 10 | 11 | 12 |
> | 4 | 4 | 5 | 6 | 7 | 8 | 9 | 10 | 11 | 12 | 13 |
> | 5 | 5 | 6 | 7 | 8 | 9 | 10 | 11 | 12 | 13 | 14 |
> | 6 | 6 | 7 | 8 | 9 | 10 | 11 | 12 | 13 | 14 | 15 |
> | 7 | 7 | 8 | 9 | 10 | 11 | 12 | 13 | 14 | 15 | 16 |
> | 8 | 8 | 9 | 10 | 11 | 12 | 13 | 14 | 15 | 16 | 17 |
> | 9 | 9 | 10 | 11 | 12 | 13 | 14 | 15 | 16 | 17 | 18 |

With this open question, students are allowed to decide what patterns to list. They can choose very simple patterns (e.g., noticing that numbers in rows or columns go up by one) or more complex ones (e.g., noticing that the southwest–northeast diagonals are all made up of the same number). Some students might even notice that there are as many 0s as 18s, 1s as 17s, 2s as 16s, and so forth. In asking for only two patterns, the question does not subject struggling students to excessive pressure. In the class conversation, of course, many more than two patterns are likely to come up.

How extensively the teacher discusses the underlying mathematics that explains each pattern depends on what the students are ready for in terms of development and background knowledge.

These pictures show what 1, 2, 3, 4, 5, and 6 look like on a die.

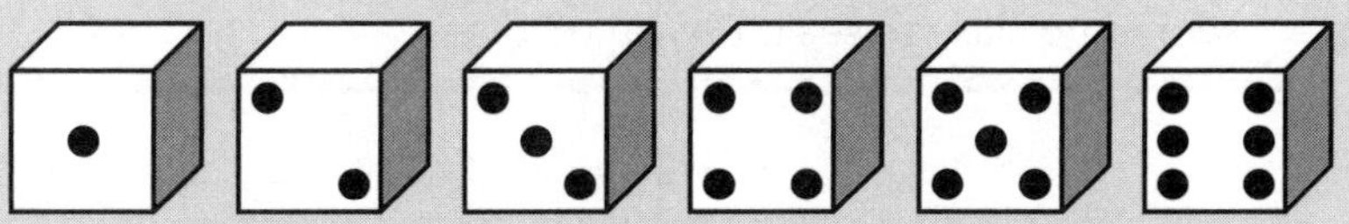

If a die had dots for 7 and for 8, what do you think the dot patterns would look like? Why?

Students might observe the pattern that is used to show 2, 3, 4, 5, and 6 and then generalize that pattern (even numbers come in pairs; odd numbers have a central dot between the pairs). Most students will find 8 easier to design than 7, but there are a number of possibilities that would make sense. A question such as this one shows students how certain arrangements make patterns easier to see than other arrangements.

TEACHING TIP. Subitizing (i.e., recognizing amounts visually without counting) has been shown to be an important number skill. Helping students recognize amounts is often based on arranging them in patterns that are easy to summarize mentally.

OPEN QUESTIONS FOR GRADES 3–5

✳ **BIG IDEA.** A group of items form a pattern only if there is an element of repetition, or regularity, that can be described with a pattern rule.

A pattern begins like this: 2, 6, . . .

How might it continue?

There are many possible responses students might come up with. A few are shown here:

2, 6, 10, 14, 18, . . .   2, 6, 11, 17, 24, . . .
2, 6, 2, 6, 2, 6, . . .   2, 6, 18, 54, . . .
2, 6, 3, 7, 4, 8, . . .

Variations. This question can easily be adapted by changing the first two numbers shown.

TEACHING TIP. Although there is just as much variability in how to continue patterns where the first three terms are given as in cases where the first two terms are given, it appears that students are much less likely to recognize the variability when the first three terms are given. For example, if they are given 1, 3, 5, most students will continue the pattern either by repeating 1, 3, 5 or by adding 2 to get 7, 9, 11. If they are given 1, 3, students seem more likely to see more alternatives.

A pattern starts with 3, and the fifth number is 11. What could the in-between numbers be?

$$3, \underline{\quad}, \underline{\quad}, \underline{\quad}, 11, \ldots$$

Most students will assume that this is a **linear pattern**, that is, one that grows by a constant amount each time. Therefore, they will assume that the missing numbers are 5, 7, and 9. However, there is potential for other responses, for example:

3, <u>11</u>, <u>3</u>, <u>3</u>, 11, 3, 3, 3, 11, . . .
3, <u>4</u>, <u>6</u>, <u>9</u>, 11, 14, 18, 21, 25, . . . (+1, +2, +3; +2, +3, +4; +3, +4, +5, . . .)
3, <u>4</u>, <u>5</u>, <u>10</u>, 11, 12, 17, 18, 19, . . . (3 consecutive numbers and then add 5)

Variations. This question can be adapted by changing the numbers of the pattern that are given or by changing the positions of those numbers within the pattern.

TEACHING TIP. One of the big mistakes commonly made in teaching patterns is expecting students to assume that there is only one way to continue a specific pattern. Unless a pattern rule is provided, this would never be true.

Use shapes like these to make a pattern that changes in two ways.

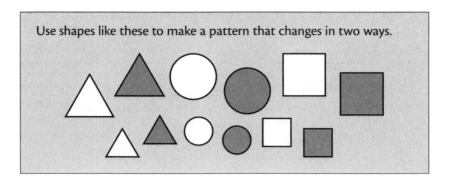

Some students find it much more difficult to deal with patterns that change in two ways than patterns that change in only one way. By leaving the question open, it can become more accessible even to struggling students. For example, one student might use a large and small triangle followed by a large and small circle and repeat that core, thus changing both shape and size; another might create a more complex pattern, such as the one below:

✸ **BIG IDEA.** **Many number, geometry, and measurement ideas are based on patterns.**

> Imagine a large group of people sitting in a circle. The first person says 1, the next one 2, and so on. But if the number they are supposed to say has a 3 in it, or if it is a multiple of 3, they say "Buzz" instead of the number and are eliminated.
>
> Choose a number of people for your circle. After several rounds of the game, who is the last person remaining?

Allowing students to choose the number in the circle makes the task more accessible to all students, for example, a student might select circles with only four, five, or six players. In a circle of four, the numbers work like this:

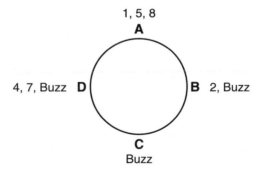

The question can also provide an appropriate challenge for those needing such a challenge; they can use circles with a larger group of people.

In follow-up discussion, the teacher could ask how someone might choose where to sit to be sure he or she is the last person left.

> You build towers of **snap cubes** to make the first five shapes in a pattern. You use between 20 and 30 cubes altogether.
>
> What could your pattern look like?

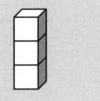

This pattern problem ties directly into arithmetic skills. Because a range of numbers of cubes is allowed, there are many possible solutions. Ideally, students will have the cubes to use, rather than simply solving the problem symbolically. One student might use towers of heights 1, 3, 5, 7, and 9; another might use towers of heights 5, 6, 5, 6, and 5; and still others might use other repeating or increasing patterns.

Variations. This question can easily be adapted by changing the number of cubes used altogether, the number of towers required, or both. Additional conditions can be put on the problem by stipulating the colors of cubes to be used.

✸ **BIG IDEA.** Arranging information in charts and tables can make patterns easier to see.

Look at the multiplication table. List two patterns you see in the table.

×	0	1	2	3	4	5	6	7	8	9
0	0	0	0	0	0	0	0	0	0	0
1	0	1	2	3	4	5	6	7	8	9
2	0	2	4	6	8	10	12	14	16	18
3	0	3	6	9	12	15	18	21	24	27
4	0	4	8	12	16	20	24	28	32	36
5	0	5	10	15	20	25	30	35	40	45
6	0	6	12	18	24	30	36	42	48	54
7	0	7	14	21	28	35	42	49	56	63
8	0	8	16	24	32	40	48	56	64	72
9	0	9	18	27	36	45	54	63	72	81

A multiplication table with a 0 row (and 0 column) was used in this example. Many teachers use this table rather than one beginning with 1.

With this open question, students are allowed to decide what patterns to list They can choose very simple patterns (e.g., noticing that numbers in rows or columns go up by the row or column header) or more complex ones (e.g., noticing that every row is either all even numbers or half even numbers). In asking for only two patterns, the question does not subject struggling students to excessive pressure. In the class conversation, many more than two patterns are likely to come up.

How extensively the teacher discusses the mathematics that explains each pattern depends on what the students are ready for in terms of development and background knowledge.

TEACHING TIP. Many patterns in tables and charts can be explained mathematically. It is valuable to let students try to come up with some of the explanations themselves.

✳ **BIG IDEA.** Variables can be used to describe relationships.

> Make up two equations that use variables and that are true all of the time. Then make up another two equations that use variables and that are true only some of the time.

It is very important that students learn the different ways equations are used. Sometimes equations are used to describe a very specific relationship: for example, $5 + n = 10$ only when $n = 5$. At other times, equations are used to describe broader relationships: for example, $n + 5 = 5 + n$ is true for all values of n and describes a basic property of numbers.

The open task suggested above allows multiple entry points. As an equation that is always true, a student might write a very simple equation, such as ▣ = ▣ or $2 \times ▣ = ▣ + ▣$, or something more complex, for example, $2 \times ▣ + 3 \times ▣ = 5 \times ▣$.

OPEN QUESTIONS FOR GRADES 6–8

✳ **BIG IDEA.** A group of items form a pattern only if there is an element of repetition, or regularity, that can be described with a pattern rule.

> A pattern is built by adding pairs of terms to get the next term. There is a 10 somewhere between the fourth term and the tenth term. What could the pattern be? Think of as many possibilities as you can.

Students are likely to work backward to solve this problem. Students who are struggling might make the 10 the fourth term. They could then recognize that

10 = 4 + 6 and make the second and third terms 4 and 6, respectively. Because the 6 comes from adding the 4 to the first term, the first term would have to be 2:

$$2, 4, 6, 10, 16, \ldots$$

If students make the 10 a higher term, they will have more work to do. For example, if the 10 were the sixth term, then they might choose the terms before the 10 to be 5 and 5:

$$\_\_\_, \_\_\_, \_\_\_, 5, 5, 10, \ldots$$

The term before the first 5 would have to be 0. The term before that would have to be 5, and so on:

$$-5, 5, 0, 5, 5, 10, \ldots$$

Variations. The question can be varied by replacing the number 10 with any other value, including fractions or mixed numbers. Also, the position of the 10 (or its replacement number) in the pattern can be altered.

✹ **BIG IDEA.** **Any pattern, algebraic expression, relationship, or equation can be represented in many ways.**

> A number pattern can be described by a **table of values**. The first number tells the position of the number in the pattern, and the second number is the pattern value. For example, the pattern 2, 4, 6, 8, . . . is described by:
>
> | 1 | 2 | 3 | 4 | 5 | 6 |
> |---|---|---|---|---|---|
> | 2 | 4 | 6 | 8 | 10 | 12 |
>
> Then the table of values can be graphed; the ***x*-coordinate** is the **position number** and the ***y*-coordinate** is the term value. Suppose the graph for a pattern goes through (2,3). What could the pattern be?

This question reinforces the concept that a pattern can be described as a list, as a table of values, or as a graph. At the same time, it provides an opportunity for many responses. For example, one student's pattern might be 3, 3, 3, 3, . . . , another might be 1, 3, 5, . . . , and yet another might be 5, 3, 1, −1, ...

TEACHING TIP. It is important that students are given experience with decreasing patterns. Instruction often focuses primarily on increasing patterns.

Draw pictures that might help someone predict the next four terms of the pattern 1, 4, 9, 16, ...

Many students will draw a representation of the pattern, but some representations may be more useful than others in predicting subsequent terms of the pattern. Two valuable visual approaches to the pattern are shown below:

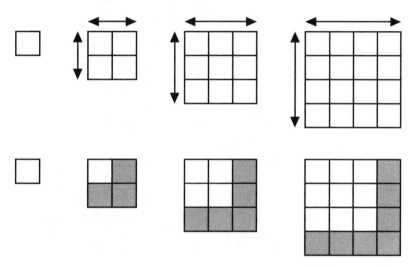

The first set of pictures emphasizes that the numbers are the products of equal values, whereas the second set of pictures emphasizes that the values increase by increasing odd numbers. It is useful for students to see how different representations of a pattern make different aspects of the pattern evident.

Variations. Another pattern that can be used to accomplish the same goal is the pattern 1, 3, 6, 10, ... A third one is a simple growing pattern such as 2, 5, 8, ... In the last case, a student might represent each term as a set of 3s with one item missing in the last group; other students might represent the pattern by emphasizing the start of 2 and a pattern based on adding 3:

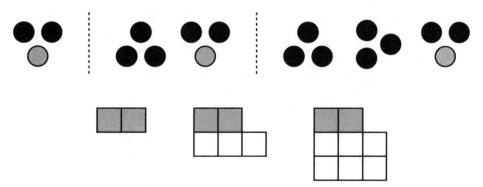

✷ **BIG IDEA.** **Many number, geometry, and measurement ideas are based on patterns.**

On a 100 chart, you color squares to form a capital letter. If you add the numbers the letter covers, the sum is between 100 and 120. What could the letter be? Where could it be?

1	2	3	4	5	6	7	8	9	10
11	12	13	14	15	16	14	18	19	20
21	22	23	24	25	26	27	28	29	30
31	32	33	34	35	36	37	38	39	40
41	42	43	44	45	46	47	48	49	50
51	52	53	54	55	56	57	58	59	60
61	62	63	64	65	66	67	68	69	70
71	72	73	74	75	76	77	78	79	80
81	82	83	84	85	86	87	88	89	90
91	92	93	94	95	96	97	98	99	100

Students can use a simple letter such as an **I** or a more complex one such as a **T**. Some students will solve the question algebraically: for example, the values within an **I** that is four spaces tall are n, $n + 10$, $n + 20$, and $n + 30$. Other students will use a more numerical technique.

As students discuss their solutions, those who did not use variables will have an opportunity to learn from those who describe how they used variables.

✷ **BIG IDEA.** **Variables can be used to describe relationships.**

Copies of three different shapes are placed in a grid. Each shape is worth a different amount. The total amounts for one of the rows and one of the columns are given. How much might each shape be worth?

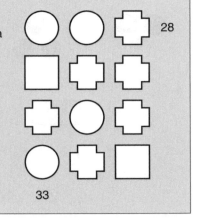

Algebraic thinking will help a student figure out why the square must be worth 5, but there is some flexibility in what the other values can be worth. For example, the circle could be worth 10 and the cross worth 8, or the circle could be worth 6 and the cross worth 16.

If all of the row and column values were given, there would be only one solution and the question would be less open. As it is, there are some constraints but some flexibility as well, making the problem open to a broader range of students.

> An expression involving the variable k has the value 10 when $k = 4$.
> What could the expression be?

This very open question allows for many responses, ranging from extremely simple answers to much more complicated ones. Possible expressions include: $k + 6$, $2k + 2$, $14 - k$, $10k ÷ 4$, $2k^2 - 5k - 2$, and so on. It would be useful for students to identify the expressions that are actually equivalent (e.g., $k + 6$ and $2 + k + 4$) as opposed to the expressions that take on the same values only sometimes (e.g., $k + 6$ and $14 - k$).

Variations. This question can easily be adapted by changing either the input or the output value, or both.

TEACHING TIP. Students are often asked to evaluate an expression for a given value of a variable. Questions in which the student is asked to create the expression for a given input/output combination are much more suitable for differentiating instruction.

> If $s = 4$ and $t = 5$, these statements are true:
>
> $$3(s + t) = 27 \qquad 2s + 3t = 23 \qquad 2t - 2s = 2$$
>
> Choose values for p and q. Write three true statements using those variables. See if a partner can figure out what your values are.

Because students can choose the values for p and q and create the equations, even struggling students will be able to participate. For example, a struggling student might choose values such as $p = 1$ and $q = 2$ and use equations such as $p + q = 3$, $q - p = 1$, and $p - 1 = 0$.

When the partner students are given the set of statements created by the first students and asked to figure out the values, the partners need to decide which statement might be the best one to start with. For example, in the set of possible statements from the struggling student given in the preceding paragraph, it is easiest to start with the last equation because it involves only one variable.

PARALLEL TASKS FOR PREKINDERGARTEN–GRADE 2

> **PARALLEL TASKS** are sets of two or more related tasks that explore the same big idea but are designed to suit the needs of students at different developmental levels. The tasks are similar enough in context that all students can participate fully in a single follow-up discussion.

✸ **BIG IDEA.** **A group of items form a pattern only if there is an element of repetition, or regularity, that can be described with a pattern rule.**

> **Option 1:** Make a pattern using blue and green snap cubes.
>
> **Option 2:** Make a pattern using blue and green snap cubes. There must be more blue cubes than green ones.

These options are very similar, although **Option 2** puts a condition on the pattern. Most young students will be able to handle either option, but some might prefer the lack of a constraint. There is a great deal of variety in terms of the patterns students can create, even for **Option 2**. For example, the pattern could begin like any of these:

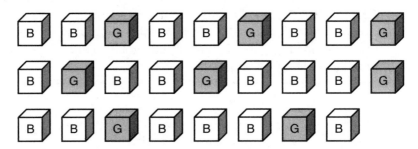

Questions the teacher could ask, regardless of the option selected, include:

- *What color is the fifth cube in your pattern?*
- *Which is the next cube that is the same color? Why is it the same color?*
- *How could you describe your pattern to someone?*

Variations. These tasks can easily be varied by using different color combinations or rules.

> **TEACHING TIP.** When numbered options are offered, sometimes the "simpler" option should be presented as Option 1 and other times as Option 2. The unpredictability will ensure that students consider both possibilities when they choose their tasks.

> ___Option 1:___ Create a repeating pattern that begins with 3, 5, ...
>
> ___Option 2:___ Create an increasing pattern that begins with 3, 5, ...

Some students are comfortable with repeating patterns but not increasing patterns; others can handle increasing patterns. By posing these parallel tasks, the teacher allows students to choose the type of pattern they wish and sets the stage for a follow-up conversation that will include both groups.

Questions such as the following would be relevant no matter which task students chose:

* *What is your pattern?*
* *What makes it a pattern?*
* *What would be your 10th number?*

Variations. These tasks can be varied by using different starting numbers or by specifying the use of a decreasing pattern rather than an increasing one.

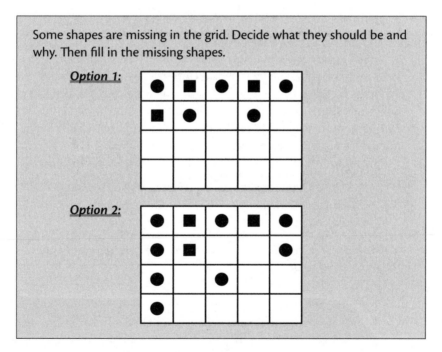

Observing patterns in two dimensions is more demanding for students than recognizing one-dimensional patterns. ___Option 2___ allows the student to focus on one dimension only, because each line can be identical, whereas ___Option 1___ does not.

___

TEACHING TIP. It is advisable to provide students with grids large enough that they can use concrete objects to make the patterns.

___

✹ **BIG IDEA.** **Many number, geometry, and measurement ideas are based on patterns.**

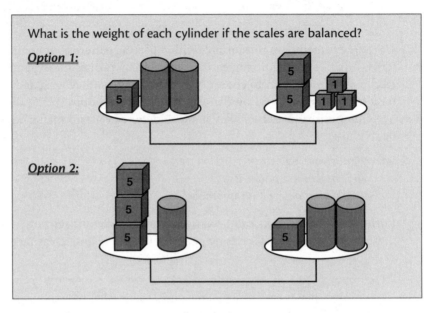

What is the weight of each cylinder if the scales are balanced?

Option 1:

Option 2:

Each option is a challenge, but a different sort of challenge. Students do need to understand that both cylinders in **Option 1** have the same weight, as do the three cylinders in **Option 2**. In **Option 1**, students need to recognize that they can mentally trade the 5 cube for five 1 cubes to figure out the value for each cylinder. In **Option 2**, students need to realize that they can remove a cube and a cylinder from both sides of the scale and balance will be preserved.

Common questions a teacher could ask all students include:

- *How did you know whether your cylinder weighed more or less than 5?*
- *What number sentence would describe the original picture?*
- *How did you figure out the weight of your cylinder?*

Variations. The numbers and weights of the cylinders and cubes on each side of the pan can be varied.

Will Rebecca and Ethan ever have the same number of stickers? How many stickers would that be?

Option 1: Ethan has 30 stickers and Rebecca has 12. Ethan gives Rebecca 3 stickers at a time.

Option 2: Ethan has 50 stickers and Rebecca has 10. Ethan gives Rebecca 5 stickers at a time.

The two proposed options are similar, each requiring both counting up and counting down. In **Option 2**, the counting is likely to be much easier for students.

Both of the problems lend themselves to students creating a pattern to arrive at a solution. For example, for *Option 2*, the pattern describing the number of stickers for Ethan is 50, 45, 40, 35, . . . and for Rebecca the pattern is 10, 15, 20, 25, ... The problem is solved by determining when two corresponding term values are the same.

Whichever task was selected, a teacher could ask:

- *How many stickers did each have after the first trade? After the second trade?*
- *How did you know that Ethan's numbers would keep going down and Rebecca's would keep going up? By how much?*

Variations. Other values can be substituted for those provided in the two options.

TEACHING TIP. A problem such as this one, which is totally suitable for pre-K–grade 2 students, would be equally suitable for much older students, who might use the equation $50 - 5x = 10 + 5x$ to solve Option 2.

PARALLEL TASKS FOR GRADES 3–5

✸ **BIG IDEA.** A group of items form a pattern only if there is an element of repetition, or regularity, that can be described with a pattern rule.

> Notice how the values inside these triangles form a pattern. The numbers inside the triangles come from adding the numbers at the corners.
>
>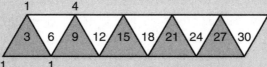
>
> **Option 1:** Figure out what numbers belong at the corners of all the triangles. What do you notice about these numbers?
>
> **Option 2:** Make your own triangle pattern where the values inside come from adding the values at the corners. Do those corner values always form a pattern?

Most students will be curious about how the numbers at the top and bottom of the triangles seem to automatically form a pattern just because the numbers inside do. By providing two options, some students will simply do the arithmetic to uncover the patterns for the top and bottom lines, whereas other students might generalize to additional situations.

Whichever option was selected, a teacher could ask:

- *What numbers did you end up with at the top? Did they form a pattern?*
- *What about the bottom? What was the pattern?*

> **Option 1:** Create three different patterns, each including the number 40 as the fourth, fifth, or sixth number in the pattern.
>
> **Option 2:** Create three different increasing patterns, each including the number 40 as the fourth, fifth, or sixth number in the pattern.

By removing a condition on the pattern that is created, a more complex task becomes accessible to a larger group of students. Students who select **Option 1** might use a combination of repeating patterns and increasing or decreasing patterns, or perhaps only repeating patterns.

Even the more complex task, **Option 2**, is fairly open. A student might use a pattern such as 10, 20, 30, 40, . . . as one of the patterns to help begin the search for other patterns that would have 40 as the fifth or sixth number. Possibilities for additional patterns include: 8, 16, 24, 32, 40, . . . and 10, 16, 22, 28, 34, 40. The student might create these by thinking about division, without or with remainders. In the case of 8, 16, 24, . . . , he or she could simply divide 40 by 5. In the case of 10, 16, 22, . . . , the student could divide 40 by 6 and realize that, because there is a remainder, if the values go up by 6, the first number cannot be a multiple of 6.

A teacher could ask any of the students:

- *How did your pattern start?*
- *How did it continue?*
- *What was your pattern rule?*
- *How did you decide how to get 40 in the fourth spot?*

Variations. These options can be adapted by changing the number 40 to a different value, changing the required positions for the number, or both.

✳ **BIG IDEA.** Any pattern, algebraic expression, relationship, or equation can be represented in many ways.

> **Option 1:** In a group of people, each person shakes hands with each other person exactly once. How many handshakes will there be if there are 3 people? 4 people? 11 people?
>
> **Option 2:** The triangular numbers are based on arranging dots in triangular patterns. How many dots will there be in the 10th picture?
>

Some students will enjoy acting out the handshake problem rather than drawing dots. Other students probably will prefer to independently draw dots or tally handshakes mathematically.

It turns out that the patterns needed to complete both tasks are identical (i.e., 1, 3, 6, 10, 15, . . .) and that the number of handshakes among 11 people is the same as the number of dots in the 10th dot picture.

Students will see how the same pattern emerges from very different situations through common discussion of the problems. The teacher could ask both groups:

- *How did you figure out your number?*
- *Do you notice a pattern? What was the pattern?*
- *What was the fourth number in the pattern? How did you figure it out?*
- *What was the sixth number?*
- *Why did the numbers in the pattern keep increasing by different amounts?*

✹ **BIG IDEA.** Many number, geometry, and measurement ideas are based on patterns.

Choose an animal type and a square grid size. Write the letters of that animal's name one at a time in the squares of the grid, going left to right, from top to bottom. For example:

M	O	N	K
E	Y	M	O
N	K	E	Y
M	O	N	K

Option 1: Predict what the last letter you write will be if the grid is a 6 by 6 grid. Test your prediction. Explain how you can predict.

Option 2: What size grid would you need to use to make sure that the first letter in the animal's name is the last one you write?

It may not be immediately apparent that both options relate to divisibility. The reason MONKEY ended with a K in the 4 by 4 grid is because there are 16 squares in the grid and the word MONKEY has 6 letters. The remainder when dividing 16 by 6 is 4, and thus it is the fourth letter in MONKEY that appears last. Students who select *Option 2* will need to generalize this idea to recognize that the remainder must be 1 when a square number is divided by the number of letters in the animal's name. So, for example, if the name has 5 letters, a 4 by 4 grid would work, as would a 6 by 6 grid. Both have 1 more square than a multiple of 5.

Both groups could be asked questions such as:

- *What animal name did you use?*
- *Why does the number of letters in your animal's name matter?*
- *How did you predict the size of the grid that would work?*
- *How would your answer change if the number of rows had to be one greater than the number of columns?*

Option 1: The sum of the first 10 terms of a pattern is a multiple of 3. What could the pattern be?

Option 2: The sum of the first 10 terms of a pattern is 195. What could the pattern be?

Each of the options requires students to create a pattern to meet a condition. For those who realize that the sum of multiples of 3 must be a multiple of 3, **Option 1** would be an easy task. They could use any multiples of 3 to form their pattern. Even students who do not recognize the simple approach could experiment to create a pattern that might work.

Option 2 has a more stringent condition. Students would likely use a combination of guess-and-test and reasoning to come up with a possible solution. For example, 6, 9, 12, 15, 18, 21, 24, 27, 30, 33 would work.

Possible prompting questions include:

- *Was the number 12 part of your pattern? How early in the pattern did it come?*
- *Was the number 200 part of your pattern? Could it have been?*
- *If all the terms were positive numbers, why would some of them probably be less than 20?*
- *What was your pattern?*
- *How did you figure it out?*

✳ **BIG IDEA.** **Variables can be used to describe relationships.**

Option 1: Choose a location on the 100 grid. Describe how you could get to the numbers that are:

     31 less   37 more   18 more

Option 2: You are on a square called s on the 100 grid. Describe these locations in terms of s:

     31 less than s 37 more than s 18 more than s

1	2	3	4	5	6	7	8	9	10
11	12	13	14	15	16	14	18	19	20
21	22	23	24	25	26	27	28	29	30
31	32	33	34	35	36	37	38	39	40
41	42	43	44	45	46	47	48	49	50
51	52	53	54	55	56	57	58	59	60
61	62	63	64	65	66	67	68	69	70
71	72	73	74	75	76	77	78	79	80
81	82	83	84	85	86	87	88	89	90
91	92	93	94	95	96	97	98	99	100

These two tasks require the same thinking, although one requires the use of a variable and the other does not. A common discussion will allow students who struggle with the use of variables to gain more insight by listening to those who are using similar reasoning but with variables.

Questions the teacher could ask both groups include:

- *How does the number change if you go to the position right above you? To the position right below you?*
- *How does the number change if you go one position to the left? One to the right?*
- *How do you get to the number that is 31 less? 37 more? 18 more?*

PARALLEL TASKS FOR GRADES 6–8

✷ **BIG IDEA.** A group of items form a pattern only if there is an element of repetition, or regularity, that can be described with a pattern rule.

> Choose two numbers. Call them *x* and *y*.
>
> **Option 1:** The pattern rule is: Start at *x* and add *y*. Can the 100th term be 900 greater than the 10th term? If so, how? If not, why not?
>
> **Option 2:** The pattern rule is: Multiply the term number by *x* and add *y*. Can the 100th term be 900 greater than the 10th term? If so, how? If not, why not?

There are two types of rules describing patterns. One is called an **iterative rule**. It tells where to start and how to continue the pattern (as in ***Option 1***). The other is a **relationship rule**. It tells how to calculate the value of a term given the term number. Much of the work of 6th grade to 8th grade is moving students from the first type of pattern rule to the second. Some students will be ready to move more quickly than others.

The question posed requires a generalization. Students must realize that in the first instance, if terms are 90 apart, and y is added each time, then $90y = 900$; thus it is 10 that must be added each time. In the second case, the 10th term is $10x + y$ and the 100th term is $100x + y$. The difference is $90x$, so $x = 10$.

Many students will reach these conclusions by using guess-and-test rather than by using algebra. In the discussion of the two problems, it will become clear that one of the two values must be 10.

Questions that could be asked of all students include:

- *What if x and y were 4 and 8; what would the 10th term be? The 100th term?*
- *What if they were 4 and 6? What if they were 4 and 10?*
- *What did you notice about your patterns?*
- *Could the 100th term be 900 greater than the 10th term? How did you figure it out?*

✳ BIG IDEA. **Many number, geometry, and measurement ideas are based on patterns.**

> **Option 1:** A cube has a side length of *n* units and is made up of smaller 1-unit cubes. The cube is painted on the outside. How many of the small cubes are painted on two faces?
>
> **Option 2:** A line of linking cubes is *n* cubes long. The line of cubes is painted on the outside. How many cubes are painted on four faces?

In each situation, students are asked to generalize a situation, which is what thinking algebraically is all about. In the first case, they must realize that the only cubes painted on two faces are along edges of the cube but not at the corners. They must then consider how many cubes are along each edge. In the second case, students must recognize that all but the end cubes are painted on four faces. In either case, students would likely, although not necessarily, use a variable to describe the number of cubes fitting the requirement.

Questions that could be asked of both groups include:

- *How many faces of a cube could be painted? Why only those numbers?*
- *What number of painted faces is most likely? Why?*
- *How can you describe your number in terms of* n?

Option 1: On a geoboard, create shapes with the indicated number of pegs on the border and inside of them and calculate their areas. What patterns do you notice?

Border	Inside	Area
3	0	
3	1	
3	2	
3	3	

Border	Inside	Area
3	1	
4	1	
5	1	
6	1	

Option 2: On a geoboard, create triangles with the indicated base and height lengths. What patterns do you notice?

Base	Height	Area
4	1	
4	2	
4	3	
4	4	

Base	Height	Area
4	2	
5	2	
6	2	
7	2	

Some students will find **Option 2** accessible because it is more obvious how to create the required shapes and easier to determine their areas.

Both of these tasks help students make sense of different formulas for area. The first formula is $A = \frac{B}{2} + I - 1$; the other is $A = bh \div 2$. Both are developed using patterns. For example, the first part of **Option 1** might be represented in this fashion:

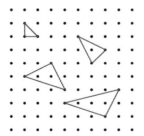

Border	Inside	Area
3	0	$\frac{1}{2}$
3	1	$1\frac{1}{2}$
3	2	$2\frac{1}{2}$
3	3	$3\frac{1}{2}$

Whichever task was chosen, a teacher could ask:

- *How do the numbers in the table change?*
- *How do the areas change?*
- *Why do you think the areas increase each time?*
- *How would changing the first value in your tables change the area?*
- *How would changing the second value in your tables change the area?*

✹ **BIG IDEA.** **Variables can be used to describe relationships.**

What is $a * b$ if the following statements are true?

Option 1	Option 2
$2 * 4 = 8$	$2 * 4 = 20$
$2 * 5 = 9$	$3 * 4 = 25$
$2 * 6 = 10$	$5 * 6 = 61$
$3 * 4 = 10$	$8 * 9 = 145$
$3 * 5 = 11$	
$3 * 6 = 12$	

These two tasks are similar in that each requires students to figure out an **operation rule** (i.e., a statement of what was done to the two given input numbers to get the output value).

Option 1 is scaffolded a bit more than **Option 2** because only one variable is changed at a time. By observing that an increase of 1 in the second value increases the result by 1, but an increase of 1 in the first value increases the result by 2, the student might realize that the first number is doubled and added to the second number. For **Option 2**, students must use what they know about number properties to figure out that both numbers are squared and added.

Common questions for both groups could include:

- *What do you think $2 * 7$ would be? Why?*
- *Do you think that $4 * 9$ is double $2 * 9$? Why or why not?*
- *How did you figure out what $4 * 9$ is?*
- *What is a $*$ b?*

Variations. The options can be varied by using simpler or more complex operation rules and more or less scaffolded examples.

Option 1: Write a word problem to match the equation $28n + 75 = 215$.

Option 2: Write a word problem to match the equation $4n + 3 = 27$.

Each task requires students to connect contextual situations with equations, and either is appropriate for students at this level. The **coefficient** and **constant** are less obvious in **Option 1**, so students might have to think harder about situations that might involve these values.

No matter which equation was used, students could be asked:

- *How does your problem use the coefficient of* n?
- *How does it use the constant?*
- *What led you to the situation you used in your problem?*

A student company charges a $5 flat fee plus $3 per window to wash windows.

Option 1: How much more would someone pay to have 35 windows washed than 24 windows?

Option 2: Might someone have to pay exactly $87 to have their windows washed? Explain.

Both of these questions require students to use the relationship described in the problem set-up. In **Option 1**, the student simply applies the rule to determine the output (price) for a given input (number of windows). This is somewhat more straightforward than **Option 2**, where the student uses either reasoning or algebra to decide whether the suggested final cost is possible. Students with good number sense may quickly realize that because 87 is a multiple of 3, it cannot be 5 more than a multiple of 3 and thus cannot be the total window washing cost.

Order these values from least to greatest. Will your order be the same no matter what the value of *n* is? Explain.

Option 1: $\frac{n}{2}$ $3n$ n^2 $3n + 1$ $10 - n$

Option 2: $4n$ $3n$ $10n$ $3n + 1$ $5n + 2$ $-n$

Some students will consider only positive whole numbers and may need prompting to consider negative numbers or fractions. In either case, students need to understand the role of a variable in describing a relationship. They can use a variety of strategies to come to a conclusion. For example, a student might use a combination of reasoning, visual representations such as a graph, and **substitution** to order their values. **Option 1** is more challenging because there is more variation in the type of relationships used.

A teacher could ask the following questions of students who worked on either task:

- *Suppose* n = 0. *Would your ordering change?*
- *Suppose* n = −1. *Would your ordering change?*
- *How do you know that* 3n < 3n + 1 *no matter what value* n *has?*

> Suppose $a@b$ means $a + ab - b/2$.
>
> **Option 1:** Is it possible for $a@b$ to be equal to 11? Explain.
>
> **Option 2:** Is it possible for $a@b = b@a$? Is it possible for $a@b = c@a$ if b and c are different? Explain.

Both problems require students to use variables, but **Option 2** requires more reasoning than **Option 1**.

A teacher could ask the following questions of students who worked on either option:

- *What would $a@1$ be?*
- *What about $1@a$?*
- *How do you know that $10@10$ is greater than 100?*
- *Which grows faster, $a@b$ or $b@a$?*

Variations. The definition of $a@b$ can be changed.

SUMMING UP

MY OWN QUESTIONS AND TASKS

Lesson Goal: Grade Level: _____

Standard(s) Addressed:

Underlying Big Idea(s):

Open Question(s):

Parallel Tasks:
Option 1:

Option 2:

Principles to Keep in Mind:
- All open questions must allow for correct responses at a variety of levels.
- Parallel tasks need to be created with variations that allow struggling students to be successful and proficient students to be challenged.
- Questions and tasks should be constructed in such a way that will allow all students to participate together in follow-up discussions.

The six big ideas that underpin work in Algebra were explored in this chapter through more than 40 examples of open questions and parallel tasks, as well as variations of them. The instructional examples provided were designed to support differentiated instruction for students at different developmental levels, targeting three separate grade bands: pre-K–grade 2, grades 3–5, and grades 6–8.

The examples presented in this chapter only scratch the surface of possible questions and tasks that can be used to differentiate instruction in Algebra. Surely many new ideas have already come to mind. For example, you might have students look for patterns in types of tables and charts differing from those shown in the examples, patterns can start with different values than were specified in the examples, or students can use alternative ways to describe relationships. A form such as the one shown here can serve as a convenient template for creating your own open questions and parallel tasks. The Appendix includes a full-size blank form and tips for using it to design customized teaching materials.

Data Analysis and Probability

DIFFERENTIATED LEARNING activities in data analysis and probability are derived from applying the NCTM process standards of problem solving, reasoning and proof, communicating, connecting, and representing to content goals of the NCTM Data Analysis and Probability Standard, including

- formulating questions that can be addressed with data, and collecting, organizing, and displaying relevant data to answer these questions
- selecting and using appropriate statistical methods to analyze data
- developing and evaluating inferences and predictions that are based on data
- understanding and applying basic concepts of probability

TOPICS

To create appropriate questions to differentiate instruction in data analysis and probability, it is important to know how students differ with respect to concepts in that strand. In terms of data display, students first use concrete graphs, then pictorial graphs where one item in the graph represents one actual item, then graphs where one item in the graph represents multiple items, and then more complex graphs such as histograms, circle graphs, and box plots. In terms of data collection and description, students move from collecting very simple yes/no data to collecting and interpreting much more complex data from either primary or secondary sources. Probability analysis usually begins with using qualitative descriptions of probability, advancing to quantitative descriptions of experimental and then theoretical probability, and sometimes on to development of models for representing probability situations.

How student understanding of these concepts develops over time can vary, of course. One useful source for a description of the typical development pattern is the NCTM Curriculum Focal Points (NCTM, 2006).

Prekindergarten–Grade 2

Within this grade band, students apply a variety of **sorting rules** to compare objects, shapes, and numbers. They also begin to use **picture graphs** and **bar graphs** to compare data. The treatment of probability is completely informal.

Grades 3–5

Within this grade band, students analyze graphs more completely, and they learn to work with more complex bar graphs and picture graphs, often to solve problems. They also might use **line plots**, **stem and leaf plots**, **double bar graphs**, and **line graphs** to display data. They begin to use coordinate grids. In many classrooms, students conduct simple experiments and describe probability in terms of what happened and what might be likely to happen in future experiments.

Grades 6–8

Within this grade band, students continue to use data displays to solve problems. New display types include **histograms** and circle graphs and sometimes other graph types such as **scatterplots** and **box and whisker plots**. They also learn to use **summary statistics**, such as **mean** or **median**, to describe a set of data.

Students in this grade band might continue to analyze simple probability situations involving single and **compound events** and determine **theoretical probabilities**.

THE BIG IDEAS FOR DATA ANALYSIS AND PROBABILITY

Coherent curricula in data analysis and probability that meet NCTM content and process standards (NCTM, 2000) and support differentiated instruction can be structured around the following big ideas:

- Many data collection activities are based on the sorting of information into meaningful categories.
- To collect useful data, it is necessary to decide, in advance, what source or collection method is appropriate and how to use that source or method effectively.
- Sometimes a large set of data can be usefully described by using a summary statistic, which is a single meaningful number that describes the entire set of data. The number might describe the values of individual pieces of data or how the data are distributed or spread out.
- Graphs are powerful data displays because they quickly reveal a great deal of information.
- An **experimental probability** is based on past events, and a theoretical probability is based on analyzing what could happen. An experimental probability approaches a theoretical one when enough random samples are used.
- In probability situations, one can never be sure what will happen next. This is different from most other mathematical situations.
- Sometimes a probability can be estimated by using an appropriate model and conducting an experiment.

The tasks set out and the questions asked while teaching data analysis and probability should be developed to reinforce these ideas. The following sections present numerous examples of application of open questions and parallel tasks in development of differentiated instruction in these big ideas across three grade bands.

CHAPTER CONTENTS			
Open Questions: Pre-K–Grade 2	151	Parallel Tasks: Pre-K–Grade 2	165
Open Questions: Grades 3–5	156	Parallel Tasks: Grades 3–5	170
Open Questions: Grades 6–8	160	Parallel Tasks: Grades 6–8	175

OPEN QUESTIONS FOR PREKINDERGARTEN–GRADE 2

OPEN QUESTIONS are broad-based questions that invite meaningful responses from students at many developmental levels.

✶ **BIG IDEA.** Many data collection activities are based on the sorting of information into meaningful categories.

> How could you sort a group of toys and make a graph to show how many toys are in the different groups?

It is a good idea for the teacher to help get this activity started by suggesting that students name some of their favorite toys and creating a list on the board. In this way, it will be easier for students to come up with ideas for how they might sort toys. Sorting criteria might be whether or not the toys are electronic, whether or not they are board games, whether younger or older children like to play with them, how many people can play with them at one time, and so on.

Students should be encouraged to make a picture graph, a bar graph, or a **concrete graph** by using cubes or some other object, depending on student familiarity with the different types of graphs. Because students are free to sort the way they wish and to graph the way they wish, this task is accessible to many students. It appeals to those who can use only simple sorting criteria and to those who can employ more complex criteria. It suits those who can create only concrete graphs as well as those who are comfortable with more abstract graphs.

TEACHING TIP. When young students are asked to collect data, it often makes sense to have them work with items in their everyday life. However, there may be times when it is beneficial to shift the focus to mathematical items, such as numbers or shapes, to integrate different mathematics strands.

[**TO THE TEACHER:** Use a group of **attribute blocks**. These blocks vary by shape, size, color, and often thickness. To begin the activity, put together a group of similar blocks, for example, all of the green triangles, whether they are large or small.]

The teacher has sorted the group of blocks in front of you. What was the sorting rule that was used? Now, sort the blocks in a different way and ask another student to figure out your sorting rule.

In this question, students again have the opportunity to create their own sorting rule. They can focus on one attribute, such as color or shape or thickness, or on more than one attribute. This flexibility allows for differentiating among students, some of whom require a simple sorting situation and some of whom do not.

Variations. This question can be altered by using other items with several attributes. For example, buttons—which can vary in size, color, shape, material, and attachment method—make excellent sorting items.

[**TO THE TEACHER:** Using sorting materials of any type, choose four items that are similar and different in various respects.]

Look at the items the teacher has placed in front of you. Which one does not belong? Explain. Think about the items in a different way. Try to find a different item that doesn't belong if you think about them in this new way.

Many students are familiar with the game "Which Doesn't Belong?" What makes this an open question is that there is usually more than one correct response, especially if the items are carefully chosen. For example, using the shapes in the figure on the next page, the most obvious response might be that the triangle does not belong because it is the only white shape. Students might also suggest, however, that the circle does not belong because it is the only round shape or that the second square does not belong because it is the only small shape.

Variations. Other sets of items (drawings or concrete items) and other numbers of items can be used to create new questions. Students can be allowed to create their own groups of items and work with partners who will be asked to find the item that does not belong.

TEACHING TIP. The data strand is ripe for differentiation. It is the *processes* of data collection, data display, and data description that are important mathematically and not the specific data themselves. There is often no reason not to let students collect data about something of personal interest to them.

✷ BIG IDEA. **To collect useful data, it is necessary to decide, in advance, what source or collection method is appropriate and how to use that source or method effectively.**

> Think of something that might be true about most of the students in the class. Conduct a survey to find out if you are correct. Display your results.

Students naturally enjoy finding out about topics of interest to them, so there is no need to define the specific topic they should pursue. Being allowed to select their own topic will enhance students' engagement with the activity. They might be interested in classmates' pets, their favorite foods, their favorite sports, how many siblings they have, or perhaps something clear-cut, such as whether or not their classmates like a certain type of animal.

With the openness of this question, the students are not only choosing the topic of their survey, but they are also deciding how to collect the data and how to display their findings. Some students might survey a sample, others the whole class. Some students might display results in a chart, others in a graph. The question offers a great deal of flexibility.

> Think about a question you might ask your classmates to which there are three possible answers. Conduct a survey and then graph the results by using either a concrete graph or a bar graph.

To collect useful data, students must learn to anticipate possible responses. This question requires such planning by the students, yet allows great flexibility. Students might, for example, ask which of three colors other students like most, which of three television shows other students prefer, or which of three places would be their favorite for a vacation. Students will quickly realize that they need to be careful in how they pose their question. For example, a student might want to ask, *Which pet do you have—a dog, a cat, or a bird?* With such a question, the student would have to be prepared for respondents who want to select two or three or no choices, as well as those who name other pets. Some students might be more comfortable rephrasing the question in a less restrictive way and coming up with their own strategies for dealing with nonstandard answers (e.g., adding an "other" bar to the graph).

To accommodate students with a wide range of graphing abilities, a choice is given for using either a concrete graph or a bar graph, depending on what is most comfortable for each individual. Linking cubes can be provided to make creating a concrete graph easier for struggling students.

Variations. Variations of this task are not difficult to create. A different number of categories can be requested, or students can be asked to collect data from someone other than classmates.

> Think of a favorite song or poem in which there are not too many words, but some words are repeated. An example is "Itsy Bitsy Spider," where the words *the, itsy, bitsy, spider, spout, rain,* and *and* are among the repeated words. Create a display to show each word in the song or poem and how many times each word appears.

Allowing students to select their own song or poem is a first step toward differentiating instruction. Further, allowing students to choose their own display method, whether, for example, a **tally chart** or a graph, is another step.

Once the displays have been created, students should have an opportunity to share their work, because doing so provides an opportunity to enhance their mathematical communication skills. For example, a teacher could ask:

* *How did you decide what kind of display to use?*
* *How did you create your display?*
* *How do you know your data are correct?*

Variations. Instead of using a poem or rhyme, students can be asked to simply select a page or a paragraph in a book and perform the same task. To keep the project manageable, students can be asked to choose a limited number of specific words to tally (ignoring the others) or to tally the most common words or the words with a specific number of letters.

Open Questions for Pre-K–Grade 2

TEACHING TIP. When a suggestion is offered in an open question, as was done by mentioning "Itsy Bitsy Spider," students may feel that they must take it. This can narrow the potential for differentiation because students feel less in control of interpreting the question in their own way. Teachers should be judicious about how often this type of scaffolding is provided, recognizing that for some students who have a difficult time getting started, a suggestion may be helpful, whereas for others it can limit expression.

✹ **BIG IDEA.** **Graphs are powerful data displays because they quickly reveal a great deal of information.**

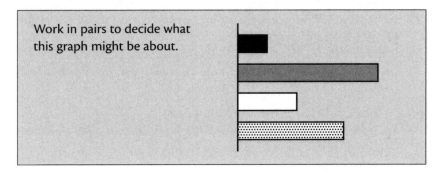

Work in pairs to decide what this graph might be about.

Because the graph is deliberately vague, offering no title or labels, students have a great deal of latitude in deciding what it might be representing. For example, one pair might suggest that it is a graph showing the favorite kind of pizza for a group of people. The students should then have to explain what each bar represents (e.g., pizza with olives, pepperoni, or extra cheese, or vegetarian pizza) and why they believe the values make sense (e.g., more people like pepperoni than anything else, very few people like olives, etc.). The students can also be encouraged to suggest the values that they think each bar might represent. Although, of course, there is no single correct answer, the teacher should steer the discussion, if necessary, to make sure that students' values make sense, for example, that the number for the longest bar is about three or four times the number for the shortest bar.

In the follow-up discussion, the teacher should draw students' attention to the variety of responses about what was represented by the graph. The fact that a graph without labels or a title could reasonably represent so many kinds of information demonstrates what powerful tools graphs can be.

TEACHING TIP. Discussion of a graph without titles and labels provides the teacher with the opportunity to help students understand the difference between a situation where many answers are desirable (e.g., an unlabeled graph) and a situation where everyone is expected to draw the same conclusion (e.g., a graph that is clearly labeled).

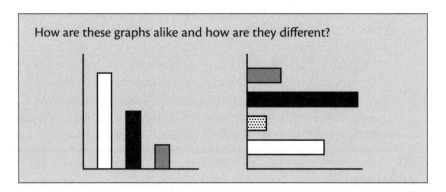

How are these graphs alike and how are they different?

Asking students to compare two items and notice similarities and differences always supports differentiation. There is deliberately no mention of what the bars represent to extend the potential of the task to meet the needs of a broad range of learners. Many students will focus on the colors of the bars, the number of bars, or the orientation of the bars as either vertical or horizontal. These are useful features to notice, but the teacher should encourage more attention to the underlying mathematics by asking questions such as:

- *How do the categories compare in size?*
- *Does one category have about twice as much as another on either graph?*
- *Can you tell how much more the biggest category is than the smallest one on either graph?*

OPEN QUESTIONS FOR GRADES 3–5

✴ **BIG IDEA.** **Many data collection activities are based on the sorting of information into meaningful categories.**

Sort the numbers from 1 to 20 by using two sorting rules so that there are four numbers that are in the overlap (that is, they fall in both categories).

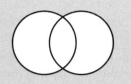

This question provides an opportunity for differentiation because students choose their own sorting rules. In the course of creating a solution to this question, all students will need to call on many aspects of what they already know about numbers. For example, a student might decide that one category is even numbers and the other category is numbers less than 10. Some students will interpret the requirement that four numbers must be in the overlap as meaning exactly four numbers are needed, and others will interpret the statement as meaning that four is the minimum number of numbers, allowing for even more options.

If students need some prompting to get them started, when the question is initially posed, suggestions for different ways to sort numbers between 1 and 20 can be solicited and shared. The sorting rules might involve attributes such as whether the numbers are in the four times table, whether they have one digit or two digits, whether they have digits that are curved when they are written as numerals, and so on.

Variations. The number of numbers required in the overlap can be changed to a different value.

✹ **BIG IDEA.** **To collect useful data, it is necessary to decide, in advance, what source or collection method is appropriate and how to use that source or method effectively.**

> Describe a situation in which you might gather data by using an experiment. Why would an experiment make sense in that situation?

Students can call on knowledge they have from their everyday lives or school knowledge to help them respond to this question. Some students might think about physical tests, for example, how many jumping jacks people of a certain age can do or how many words they can write in 2 minutes; others might think of experiments about whether different types of medicines work or experiments involving growing plants under different conditions. Although it is possible to begin the discussion with some suggestions of possible experiments, the results might be more interesting—and the question more open—if students are allowed to come up with their own ideas first.

Variations. Instead of focusing on data collected through experiments, the question can be changed around to focus on collecting data through observation or a survey, asking about the types of observations or survey questions needed to gather the required information.

✹ **BIG IDEA.** **Graphs are powerful data displays because they quickly reveal a great deal of information.**

> Select a graph type you would use to display these data. Why is your choice a good way to show the data?

Favorite Dinosaurs	
Tyrannosaurus Rex	25
Triceratops	3
Stegosaurus	8
Brachiosaurus	2

It is important for students to learn that different types of graphs are appropriate in different situations. By allowing students to choose their own graph type, instead of telling them what graph to use, it is likely that they will think more deeply about the reasons behind using different types of graphs in different situations.

Open Questions for Grades 3–5

In addition, allowing students to choose the type of graph to use provides opportunities for students with different levels of knowledge to succeed at the task. If, for example, students are asked to create a particular type of graph, those who are less comfortable with that graph type will have less chance for success. With the more open approach, as other students describe their work, even struggling students will gain experience with graph types they might otherwise avoid.

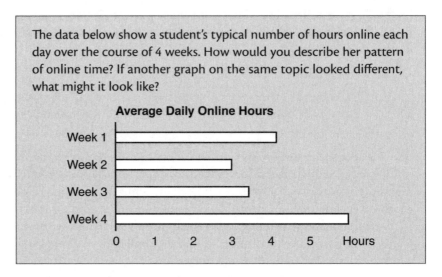

The data below show a student's typical number of hours online each day over the course of 4 weeks. How would you describe her pattern of online time? If another graph on the same topic looked different, what might it look like?

Average Daily Online Hours

This question is open in that how much students describe or what they focus on is purely up to them. One student might focus on how the values increase or decrease over the 4 weeks, another on the fact that the time spent online is considerable for certain weeks, and others on the fact that the values are all greater or less than a certain amount.

In creating a graph that looks different, students have many options. Some students might simply use a different type of graph (e.g., a **pictograph** or line plot) to display the same data, or they might use a similar graph but with a different **scale** or a different orientation. Other students might generate data they see as substantively different from the data provided, and then present the new data in a graph similar to the one shown or in a different format.

Variations. Different data or a different type of graph can be provided. Students can also be instructed to create a graph that is similar to the given one, rather than different from it.

A pictograph has been created, and these two symbols appear somewhere on the graph. Describe as much as you can about what each symbol probably represents. Explain your thinking.

Learning to use a scale indicator in a graph is a big step for this grade band. Students need to recognize that the use of partial symbols is often necessary when presenting a scale with a pictograph. In this case, students can use visual skills to surmise that the scale is probably a multiple of 4 because a quarter symbol and a half symbol were used. Some students will assume that the scale must be 4; others will realize it could be a multiple of 4. Some more advanced students may realize that there is no way to know. For example, if a whole circle represents 3, the half symbol could represent half that amount and the quarter symbol one fourth of that amount, assuming that the data represent fractions and not whole numbers.

TEACHING TIP. Many students struggle with pictograph symbols involved in scales. For example, rather than using a circle to represent four items, they might use a shape with four marked sections if they are uncomfortable using a single item to represent 4.

✸ **BIG IDEA.** An experimental probability is based on past events, but a theoretical probability is based on analyzing what could happen. An experimental probability approaches a theoretical one when enough random samples are used.

> Imagine that a new student is about to join the class. Decide which of these statements is likely, which is certain, which is unlikely, and which is impossible.
>
> - The student is a boy.
> - The student is the same age as many other students in the class.
> - The student is 20 years old.
> - The student has a head.
> - The student lives in the local area.
> - The student likes school.
> - The student has very few close friends.
>
> Add your own statements that other students can judge as likely, unlikely, certain, or impossible.

When first learning about probability, students are much more successful using qualitative terms such as *likely, unlikely, certain,* and *impossible* rather than numerical ratios or fractions. The task posed in this question is accessible to many students because it is grounded in their everyday experience. It also allows for differentiation because the students are asked to make up their own statements. The statements they make up could have to do with age, gender, habits, preferences, family relationships, and so on.

Open Questions for Grades 3–5

Variations. The task can be varied by offering statements about a sports figure or musical artist, but care must be taken to not make the question inaccessible to some students by requiring too much specialized knowledge.

> Describe an event that fits each word:
>
>   Impossible   Certain   Unlikely

With such a general question, students have a lot of opportunity to be successful. They can draw on personal, everyday experience or school knowledge. For example, a student might say that it is certain that an even number does not end in 3, or that it is unlikely that a number less than 100 describes the number of people in a town.

OPEN QUESTIONS FOR GRADES 6–8

✷ **BIG IDEA.** **Many data collection activities are based on the sorting of information into meaningful categories.**

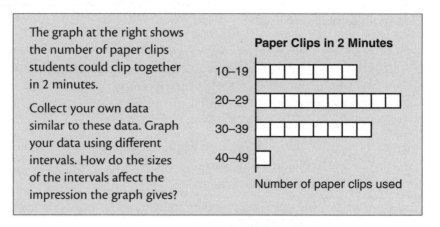

The intervals that are used to categorize data affect how a data display looks. For this question, students have a choice of exactly what data to collect; for example, they might gather data about something else students can do in 2 minutes, such as the number of times they can write their name. Then they get to choose at least two (or more, if they wish) sets of intervals with which to categorize the data. These elements of choice make the task very open. A group discussion of what different students did could, indeed, be a rich one, because of the amount of diversity possible. They also begin to work with more abstract numbers or notation, including expressions involving negative integers, exponents, and scientific notation.

✴ **BIG IDEA.** Sometimes a large set of data can be usefully described by using a summary statistic, which is a single meaningful number that describes the entire set of data. The number might describe the values of individual pieces of data or how the data are distributed or spread out.

> The mean of a set of numbers is 8. What might the numbers be?

This is a very open question. Students are free to select how many numbers to use in their set as well as the size of the numbers. Struggling students might pick a very simple set such as {7, 8, 9} or even the single number 8. But other students might select much larger data sets with more diverse values.

By providing sets of linking cubes, the teacher can make the task accessible to all students. Students might begin with any number of groups of 8 and simply move cubes around to discover new sets with the required mean. In the example shown below, the original set of {8, 8, 8, 8} cubes has been rearranged to create the new set of {7, 10, 10, 5} (mean 8).

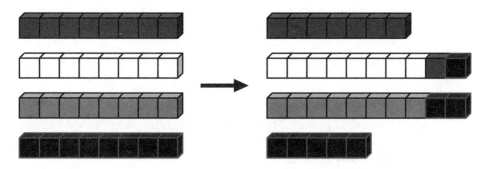

In class discussion of the different data sets students generate, the teacher can point out how, in each case, the sum of the data values is a multiple of 8. In fact, the multiple of 8 that the data values total equals the number of values in the set. For example, the sum for {7, 8, 9} is 24, which is 3 × 8 (i.e., the number of values in the set × the mean value for the set).

TEACHING TIP. Use of manipulatives is good for all students and can provide valuable scaffolding for differentiated instruction.

> Create a set of data in which the mean is greater than the median.

Once students have become familiar with concepts of both the mean and the median, they can handle a task such as this one. The question is open in that it allows students to operate at their own comfort and knowledge levels to create either simple or complex sets of data. Yet it still requires that all students deal with the concepts that a median is the middle piece of data, that a mean is a description of how data totals can be distributed evenly, and that a mean is more likely to exceed the median if there are only one or two pieces of data with high values.

To very simply perform the task, a student might begin with a set of data such as {4, 5, 6}, for which the mean is equal to the median, and then substitute a number greater than 6 for the 6.

> Create a sentence that correctly uses the following words and numbers:
>
> *average, 8, share, 2*

The exercise of creating a sentence that uses certain words allows for a great deal of student choice, a critical feature of differentiated instruction. For this task, a student might create a very simple definitional sentence such as *To get the average of 2 and 8, you have to total the two amounts and share them fairly*. A student with a greater comfort level with calculation might offer something like this: *If 2 people share the average of 8 and 10, they each get 4.5*.

As students discuss their sentences, there will be many unique contributions, each of which furthers the other students' understanding of the concept of average.

✳ **BIG IDEA.** An experimental probability is based on past events, but a theoretical probability is based on analyzing what could happen. An experimental probability approaches a theoretical one when enough random samples are used.

> A **tree diagram** showing the outcome of an experiment has 12 branches when it is complete. To what problem might the tree diagram represent the solution?

Naturally, this question is only appropriate once students have become familiar with tree diagrams. With that knowledge base in place, the question is open because the complexity of the tree diagram that is considered is up to each individual student.

Some students will draw a tree diagram representing a simple experiment, for example, spinning a spinner with 12 sections labeled **A–L** and letting each branch of the tree diagram represent one of those 12 outcomes. Other students will think of 12 as the product of 2 and 6 and use a tree diagram to represent an experiment involving two aspects, one of which has 2 possibilities and one of which has 6 possibilities, for example, flipping a coin and rolling a die.

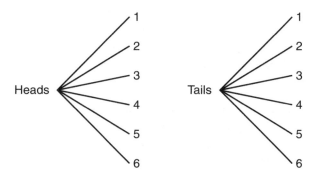

Still other students will think of a combination of two events with 3 and 4 possible outcomes, or a combination of three events with 2, 2, and 3 possible outcomes, for example, flipping two coins and spinning a spinner with three equal sections.

The discussion about the diverse tree diagrams and problems created by the students will be a rich one and will benefit a broad range of individuals.

Variations. The number of branches can be varied, although it is helpful to ensure that it is a number with several factors.

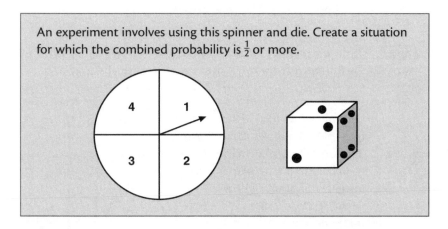

An experiment involves using this spinner and die. Create a situation for which the combined probability is $\frac{1}{2}$ or more.

This question is open in that students have a fair bit of latitude in what the probability could be. They also have freedom in deciding what the events might be. For example, a student might decide that the spinner could land on 2 or 3 (which has a probability of $\frac{1}{2}$) along with any value on the die. The combined probability would be $\frac{1}{2}$ (i.e., $\frac{1}{2} \times 1$), and the problem would be solved. Another student might decide that the combined probability of landing on a number less than 4 on the spinner and a number less than 6 on a die (i.e., $\frac{3}{4} \times \frac{5}{6} = \frac{15}{24}$) is a possible solution. A very simple solution would be allowing for any roll and any spin, for a combined probability of 1, which does satisfy the problem requirements.

Open Questions for Grades 6–8

> Imagine that you are selecting colored cubes out of a bag. What colors of cubes and how many of each color might be in the bag if the probability of selecting a green one is close to $\frac{1}{3}$?

Students have a great deal of freedom to determine the various colors of cubes in the bag as well as the numbers of cubes. The response could range from something as simple as a bag with three cubes, where only one is green, to a situation in which there are 100 cubes, 30 are green, and the rest are distributed evenly among four other colors.

In addition, students are allowed to decide what "close to" means in this context. Some might argue that a 1 in 4 probability is close enough, whereas other students might demand a fraction that is closer to 1 in 3. The mathematical concept of estimation is an important one for all students. Class discussion of this task will help students understand that estimation is not a science, but an art.

TEACHING TIP. There can never be too much discussion about estimation. It is important for students to become comfortable with deciding when to estimate and what ranges of estimates are appropriate in different situations.

✹ **BIG IDEA.** In probability situations, one can never be sure what will happen next. This is different from most other mathematical situations.

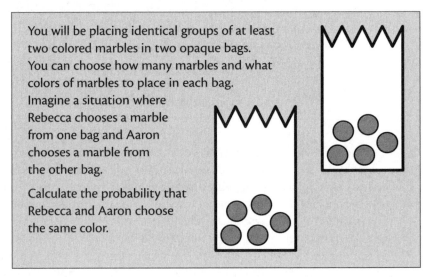

> You will be placing identical groups of at least two colored marbles in two opaque bags. You can choose how many marbles and what colors of marbles to place in each bag.
>
> Imagine a situation where Rebecca chooses a marble from one bag and Aaron chooses a marble from the other bag.
>
> Calculate the probability that Rebecca and Aaron choose the same color.

Students select the number of marbles in each bag as well as their colors. The response might be as simple as placing two blue marbles in each bag; then the probability of both students choosing the same color would be 100%. Many students, however, will select marbles of different colors. For example, if 2 green and 3 blue marbles were placed in each bag, the probability of selecting two marbles of the same color would be $\frac{13}{25}$ (i.e., the probability of selecting two greens, $\frac{2}{5} \times \frac{2}{5} = \frac{4}{25}$, plus the probability of selecting two blues, $\frac{3}{5} \times \frac{3}{5} = \frac{9}{25}$).

TEACHING TIP. When a sufficiently open question is asked, the discussion will be a rich one because there will be so many directions in which students might go and so many possible results to discuss. Inevitably, students will learn more about the concept at hand than if a more closed question is asked.

PARALLEL TASKS FOR PREKINDERGARTEN–GRADE 2

PARALLEL TASKS are sets of two or more related tasks that explore the same big idea but are designed to suit the needs of students at different developmental levels. The tasks are similar enough in context that all students can participate fully in a single follow-up discussion.

✳ **BIG IDEA.** Many data collection activities are based on the sorting of information into meaningful categories.

[**TO THE TEACHER:** Provide a set of sortable material. It might be attribute blocks; it might be buttons; it might be pictures of various colors and types of flowers.]

The teacher has set out some items that differ in various ways.

Option 1: Sort the items. Describe your sorting rule.

Option 2: Sort the items into two groups so that one group has two more items in it than the other group. Describe your sorting rule or rules.

These tasks differ in that students can use any sorting rule at all in the first instance, but in the second instance, they must use a rule that leads to a certain result. For example, if four green and two blue attribute blocks were provided, students could sort by color and solve either task correctly. If there were four green and four blue blocks, the students who selected *Option 2* would need to determine a sorting rule involving something other than color.

Whichever option was chosen, in follow-up discussion students could be asked:

- *How did you sort your items?*
- *Why was your sorting rule an appropriate one for these items?*
- *Where would this object go (as another object is held up) if we used your sorting rule?*
- *How many items fit your sorting rule(s)?*

TEACHING TIP. When numbered options are offered, sometimes the "simpler" option should be presented as Option 1 and other times as Option 2. The unpredictability will ensure that students consider both possibilities when they choose their tasks.

[**TO THE TEACHER:** Set out some items that are sortable, for example, different types of postage stamps.

Sort the items the teacher has provided. Create a bar graph to describe the number in each group after you have sorted them.

Option 1: Choose a way to sort the items so that one bar of your graph is much longer than all of the other bars.

Option 2: Choose a way to sort the items so that the bars are all about the same size.

Students choosing **Option 1** will need to find and graph something that is true about most, but not all, of the items. Those choosing **Option 2** will need to determine a way to divide the items into almost equal groups, a task which may or may not be more difficult, depending on the provided material. For example, if the collection were mostly 42¢ stamps with a few stamps worth more or less, it might be easier for a student to choose **Option 1**. If the collection were almost equal numbers of stamps with flags and stamps with flowers, it might be easier for a student to choose **Option 2**. Students are likely to use mathematical reasoning in deciding which task to complete.

Whichever option was chosen, students could be asked:

- *How did you sort your items?*
- *Why was your sorting rule an appropriate one for these items?*
- *Where would this object go (as another object is held up) if we used your sorting rule?*
- *How does your graph describe the items?*
- *What can you tell about the number of different types of items by looking at the graph?*

✳ **BIG IDEA.** Graphs are powerful data displays because they quickly reveal a great deal of information.

Option 1:
How many students chose each flavor?

Option 2:
How many students does this graph describe?

Favorite Flavor

Vanilla ▯▯▯▯
Chocolate ▯▯▯▯▯▯▯▯
Other ▯▯▯

Number of students

Some students will be able to read individual bits of information from the graph but will have more difficulty drawing conclusions from it, as is required to answer **Option 2**.

Whichever option students chose, a teacher could to ask:

- *What is this graph all about?*
- *How do you know that more students chose chocolate than vanilla?*
- *How do you know that more than nine students chose a favorite flavor?*
- *What other information does the graph tell you?*

Variations. Any graph can be substituted that is conducive to creation of one option that has students read information directly from the graph and another option that has students infer information.

What might this graph be about? Why do you think that?

Option 1:

Option 2:

Although students are taught how important it is to label a graph and provide titles, it can be very useful on occasion to provide an unlabeled, untitled graph for consideration. Deciding for themselves what the graph might be about, and hearing the diverse opinions of other students during class discussion, will help students understand why labels and titles are important.

For **Option 1**, students need to think about a situation involving two groups that are quite different in size. For **Option 2**, students need to think about a situation involving two groups that are about the same size. Students are likely to, and should, consider everyday contexts where these situations might occur.

Questions appropriate for both groups could include:

- *What title would you give the graph? Why?*
- *What labels that are missing now should be on the graph?*
- *What information does the graph tell you?*
- *Why do you think that your suggestions for the title and labels for this graph make sense?*

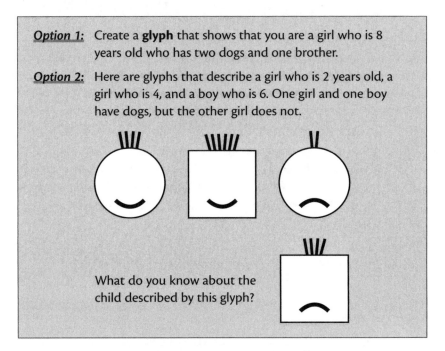

Option 1: Create a **glyph** that shows that you are a girl who is 8 years old who has two dogs and one brother.

Option 2: Here are glyphs that describe a girl who is 2 years old, a girl who is 4, and a boy who is 6. One girl and one boy have dogs, but the other girl does not.

What do you know about the child described by this glyph?

A glyph is an interesting way to show information. There are no rules about how to create glyphs, but the intent is that a pictorial code is used to convey information. For example, in **_Option 2_**, students must use reasoning to figure out the code: the shape of the face tells the gender, the number of hairs on the head tells the age, and the smile (or frown) tells whether or not the child has a dog. Once students have become familiar with the notion of a glyph after having worked with a few where the codes were explicitly given, they can select either **_Option 1_**, where they are free to create the code, or **_Option 2_**, where they are solving a problem in deciphering the code.

Questions that could be asked of students who chose either option include:

- *What are the rules for your glyph?*
- *How many rules did you need? Why that many?*
- *What do the rules tell about the picture you drew?*
- *How else might the same information have been shown using a different set of rules?*

✹ **BIG IDEA.** In probability situations, one can never be sure what will happen
next. This is different from most other mathematical situations.

Option 1: There are two bags in front of you. One of them has 8
yellow cubes and 2 blue cubes. The other one has 8 blue
cubes and 2 yellow cubes.

You pick one cube from one of the bags and it is blue. You
return the cube and pick again from the same bag. The
second cube is blue, and the one after that is yellow.

Do you think your bag has 8 yellow and 2 blue or 2 yellow
and 8 blue? Explain.

Option 2: There are three bags in front of you. One of them has 8
yellow cubes and 2 blue cubes. The second one has 5 blue
cubes and 5 yellow cubes. The last one has 8 blue cubes
and 2 yellow cubes.

You pick one cube from one of the bags and it is blue. You
return the cube and pick again from the same bag. The
second cube is blue, and the one after that is yellow.

Do you think your bag has 8 yellow and 2 blue, 5 yellow
and 5 blue, or 2 yellow and 8 blue cubes? Explain.

In both instances, students cannot be sure of their answer, but they may be
more certain in one case than the other. For students who are less comfortable
taking risks, ***Option 1*** might be a more attractive task.

All students could be asked:

- *What color do you think would be picked on a fourth try?*
- *Why do you think that?*
- *Can you be sure?*
- *Suppose you got to choose what colors to put into two or three bags. What
  colors would you put in the bags so that you could be sure which bag was
  which once you saw one or two picks? Are there other possible answers, too?*

PARALLEL TASKS FOR GRADES 3–5

✳ **BIG IDEA.** Sometimes a large set of data can be usefully described by using a summary statistic, which is a single meaningful number that describes the entire set of data. The number might describe the values of individual pieces of data or how the data are distributed or spread out.

> **Option 1:** Create a set of five pieces of data with a mean of 6. No more than one of the values can be 6.
>
> **Option 2:** Create a set of five pieces of data including the values 4, 2, and 2, and where the mean is 20.

Both tasks require that students understand the concept of mean. **Option 1** is less restrictive. Students can use any values they wish to create the data set. A student with a good understanding of the concept of mean might begin with {6, 6, 6, 6, 6} and increase two of the values and decrease two others in the same way. This approach might yield a solution of {2, 4, 6, 8, 10} (increasing and decreasing values by 2 and 4). **Option 2** requires the student to recognize the need for greater numbers in the remainder of the data set. He or she might recognize that the five values must total 100 (i.e., 5 values × 20 for the mean) and thus that the two remaining values must total 92 (i.e., 100 − 4 − 2 − 2). Another student might employ an alternate strategy.

No matter which option or approach was chosen, students could be asked:

- *What is a mean?*
- *How do you know that your mean is correct?*
- *How did you figure out the other numbers?*

TEACHING TIP. This question was set up to require working backward. Rather than giving data and asking for the summary statistic, the statistic is provided and the data are requested. This approach results in a more open question.

✳ **BIG IDEA.** Graphs are powerful data displays because they quickly reveal a great deal of information.

> The set of data below describes the ages of a group of people at a family party.
>
> 32, 30, 5, 2, 1, 62, 58, 28, 26, 25, 24, 2, 4, 39, 16
>
> **Option 1:** Create a line plot to display the data.
>
> **Option 2:** Create a bar graph to display the data.

Both of the suggested graph types are suitable for displaying the data given. A student is more likely to use **intervals** for a bar graph than for a line plot, and thus *Option 2* may be perceived to be the more difficult task (to figure out what intervals to use). On the other hand, it may be perceived to be simpler than creating a line plot because bar graphs are more familiar to most students.

Whichever task was chosen, all students could be asked:

- *How does your graph show how many people at the party are under 20? Between 30 and 40?*
- *How does your graph show the range of ages at the party?*
- *Why is your graph an appropriate way to show the data?*

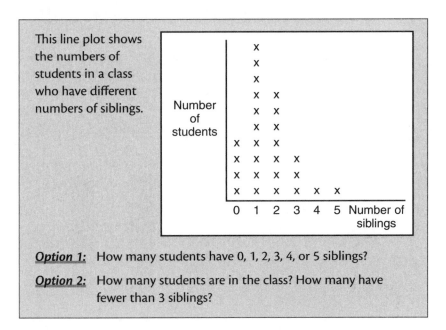

This line plot shows the numbers of students in a class who have different numbers of siblings.

Option 1: How many students have 0, 1, 2, 3, 4, or 5 siblings?

Option 2: How many students are in the class? How many have fewer than 3 siblings?

Both of the tasks require students to understand what a line plot is and how it works. *Option 1* asks students to read information directly off of the line plot, whereas *Option 2* requires them to infer information by recognizing which categories are relevant and how they need to be combined.

Students completing either task could be asked questions such as:

- *What is this line plot all about?*
- *Were there more students with 0 siblings or 1 sibling?*
- *Do you think that you would get the same data if you asked students in a different class?*
- *How many students had 0, 1, 2, 3, 4, or 5 siblings?*
- *Did any students have 6 siblings?*
- *How do you know that there were more than 10 students who answered the question?*

The final question may be somewhat more difficult than the other questions for students who have difficulty inferring information. However, it is probably simple enough (because there were 10 students in one category) that even they will be able to figure it out, and the process of working out the answer may help move them along the path toward making inferences more easily.

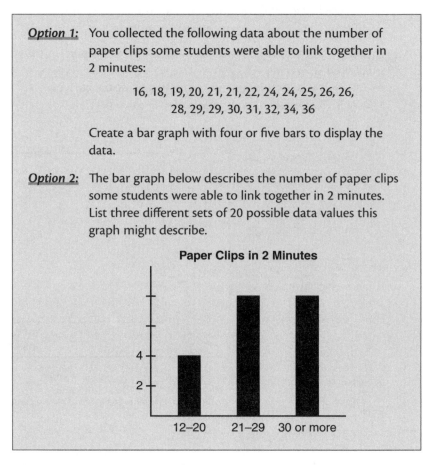

Option 1: You collected the following data about the number of paper clips some students were able to link together in 2 minutes:

16, 18, 19, 20, 21, 21, 22, 24, 24, 25, 26, 26, 28, 29, 29, 30, 31, 32, 34, 36

Create a bar graph with four or five bars to display the data.

Option 2: The bar graph below describes the number of paper clips some students were able to link together in 2 minutes. List three different sets of 20 possible data values this graph might describe.

Paper Clips in 2 Minutes

12–20   21–29   30 or more

Both of these tasks require students to work with bar graphs with intervals. In one case students must create such a graph, and in the other case they must interpret a graph. By requiring four or five intervals in *Option 1*, rather than only three, students cannot simply rely on place value concepts to define their intervals. *Option 2* emphasizes the notion that once data have been grouped into intervals, it is no longer possible to know what the exact original data were.

Students who chose either option could be asked:

- *Why do you think someone might use graphs that are based on intervals instead of having separate bars for each value?*
- *Are the intervals for your graph the same size? Why is that a good idea?*
- *Which interval had the greatest number of values in it?*
- *How does the graph show this?*
- *If different intervals had been used, would the graph have looked the same?*

✳ **BIG IDEA.** An experimental probability is based on past events, but a theoretical probability is based on analyzing what could happen. An experimental probability approaches a theoretical one when enough random samples are used.

> _**Option 1:**_ Describe what 10 colored cubes you would put in a bag so that the probability of selecting a red one is high but not certain.
>
> _**Option 2:**_ Describe what 10 colored cubes you would put in a bag so that the probability of selecting a red one is $\frac{2}{5}$.

Before students are comfortable using fractions to describe probability, they are comfortable using qualitative language to describe it (words such as "probably" or a "high probability"). With two options, students who are not ready to use fractions to describe probability can be successful, and students who can use fractions have the opportunity to do so.

In some ways, _**Option 1**_ can promote richer conversation, even though it may seem to be the simpler task. Students might debate how high is high enough. For example, could there be 8 reds and 2 other colors, or must there be 9 reds? Could there even be 6 reds? Is any probability greater than $\frac{1}{2}$ considered high?

Whichever task was chosen, students could be asked:

- _How many red cubes did you put in your bag?_
- _Why did you choose that number?_
- _Could you have chosen a different number of red cubes?_
- _What colors did the other cubes have to be?_
- _Do the other colors matter?_

Variations. The task can be adapted by using an alternate qualitative description of a probability for one task and a quantitative description of a probability for the other task. For example, Option 1 might ask students to fill a bag so that the probability of picking a blue cube is a little higher than the probability of choosing a red cube, while Option 2 might ask that the probability of picking a blue cube be $\frac{1}{8}$ more than the probability of choosing a red cube.

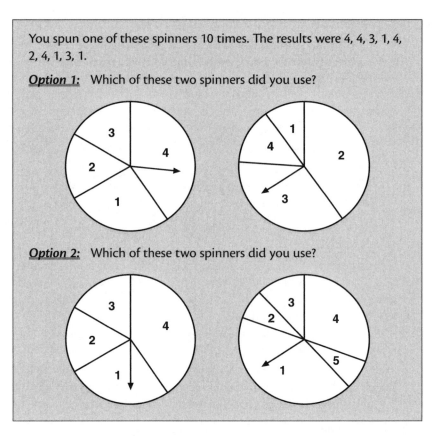

You spun one of these spinners 10 times. The results were 4, 4, 3, 1, 4, 2, 4, 1, 3, 1.

Option 1: Which of these two spinners did you use?

Option 2: Which of these two spinners did you use?

The choice about which spinner is the likely one is somewhat more obvious in **Option 1** than in **Option 2**. However, in either case, students must evaluate what happened and recognize that outcomes are not completely predictable, even though they tend to mirror theoretical probabilities to a certain extent.

In the case of **Option 1**, it seems highly unlikely that the two most frequent outcomes, 1 and 4, would occur if those were the two smallest sections. It is not certain, however, and students need to realize this. The decision is more difficult in **Option 2**. In both cases, the 1 and 4 sections are the biggest ones, as might be anticipated. Even though the fact that there was no result of 5 might dissuade a student from choosing the right-hand spinner in **Option 2**, the fact that the 2 section of the right spinner is smaller than the 2 section on the left might make them reconsider.

Students who completed either task could be asked:

* *Which outcomes influenced you the most in deciding which spinner you used?*
* *Can you be sure you are right? Why or why not?*

TEACHING TIP. Students' reasoning skills are fine-tuned more by providing them a choice of which spinner they might have used than by simply asking whether the results make sense for a particular spinner.

PARALLEL TASKS FOR GRADES 6–8

✹ **BIG IDEA.** Sometimes a large set of data can be usefully described by using a summary statistic, which is a single meaningful number that describes the entire set of data. The number might describe the values of individual pieces of data or how the data are distributed or spread out.

> **_Option 1:_** A set of nine pieces of data, all of which are different, has a mean of 30 and a median of 10. What could the data values be?
>
> **_Option 2:_** A set of nine pieces of data, all of which are different, has a median of 10. What could the data values be?

For **_Option 2_**, students need only to realize that the median is the middle number of the data set. They could simply write down the number 10 and list four different numbers less than 10 and four different numbers greater than 10. For **_Option 1_**, the student must negotiate two pieces of information. Not only does the median have to be 10, but the mean has to be 30. The student might realize that the sum of the data values must be 270 (i.e., 9 values × 30 for the mean) or might simply guess and test to find values that work.

In a discussion about the tasks, all students will demonstrate their understanding of the concept of median, but even the students who chose **_Option 2_** will be reminded about how means are calculated. Students who completed either task could be asked:

- *How do you know your median is 10?*
- *What is the smallest value you had? Did it have to be less than 10? Why?*
- *What was the greatest value you had? Did it have to be greater than 10?*
- *How did you determine your set of data values?*

✹ **BIG IDEA.** Graphs are powerful data displays because they quickly reveal a great deal of information.

> **_Option 1:_** Create a set of data that can be appropriately described by using a histogram. Create that histogram.
>
> **_Option 2:_** Create a set of data that can be appropriately described by using a stem and leaf plot. Describe that plot.

Both options require students to consider a type of graph and when it is best used. In the case of the histogram (**_Option 1_**), the student needs to recognize that the data should describe a continuous measure where it makes sense to count

Parallel Tasks for Grades 6–8

frequencies of data in different intervals. For example, the data could be heights of students in different age groups. In the case of the stem and leaf plot (***Option 2***), the student needs to recognize that it should make sense to organize the data by using place value groupings. For example, the data could be the ages of people in a race. A stem and leaf plot could even be used for data expressed as decimals, but most students are likely to use only whole-number data.

Regardless of which option a student chose, a teacher could ask:

- *What are the groupings for the data that you used?*
- *Why did it make sense to use that type of graph for these data?*
- *Looking at the graph, what picture of the data does it give the viewer?*

✳ **BIG IDEA.** **An experimental probability is based on past events, but a theoretical probability is based on analyzing what could happen. An experimental probability approaches a theoretical one when enough random samples are used.**

> The probability of an event occurring is $\frac{1}{3}$.
>
> ***Option 1:*** If you used a die, a spinner, or some other probability device, what could the event be and what might the device look like?
>
> ***Option 2:*** If you used two items in any combination of dice, spinners, or other probability devices, what could the event be and what might the devices look like?

The difference between ***Options 1*** and **2** is whether students can handle only simple events or compound events (i.e., events involving two separate aspects). For example, for ***Option 1***, students might simply choose a conventional six-sided die and suggest that the event is rolling a number less than 3; alternately, they might choose a spinner with three equal sections, one of which is red, and the event is spinning a red. For ***Option 2***, students must consider combinations. They might begin by thinking about how to write the fraction $\frac{1}{3}$ as a product of two other fractions that are less than 1; these two fractions would then help determine the two devices needed. For example, the two fractions might be $\frac{1}{2}$ and $\frac{2}{3}$, leading the student to choose two spinners, one with two equal sections labeled 1 and 2 and the other with three equal sections labeled 1, 2, and 3, and an event of getting an odd number on both spinners.

Whichever option was selected, a teacher could ask:

- *Is a probability of $\frac{1}{3}$ likely or unlikely? How do you know?*
- *What device did you choose? What probabilities are easy to describe with that device?*
- *How did you decide on your event?*
- *How do you know that the probability of your event is actually $\frac{1}{3}$?*

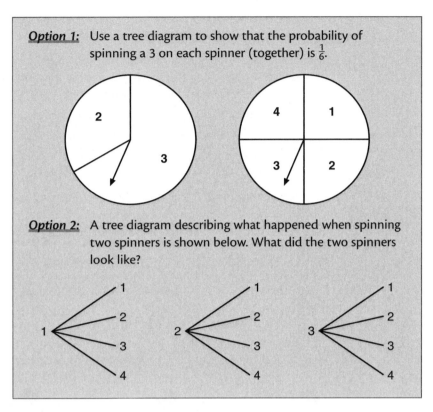

Option 1: Use a tree diagram to show that the probability of spinning a 3 on each spinner (together) is $\frac{1}{6}$.

Option 2: A tree diagram describing what happened when spinning two spinners is shown below. What did the two spinners look like?

Parallel Tasks for Grades 6–8

Both tasks require students to be familiar with tree diagrams. In **Option 1**, students must create a tree diagram from a situation. This may be a familiar problem to students, but it is a bit more challenging than the simplest cases they have encountered because the sections of one of the spinners are not of equal size. They will need to determine a way to represent this circumstance in their tree diagram. Although **Option 2** is a less traditional type of question, the presence of the spinners in **Option 1** will likely help students figure out what they might do in **Option 2**.

Whichever option was selected, a teacher could ask:

- *Why are there two "columns" in the tree diagram?*
- *What do the numbers in the first column represent?*
- *How about the numbers in the second column?*
- *How many branches are there?*
- *Why does the number of branches make sense?*

TEACHING TIP. Creating a situation and then posing a traditional question as one option and a possible answer to a similar question as another option often provides good choices for parallel tasks. Each option may serve to provide an element of scaffolding for the alternative option.

> A bag contains 3 red cubes, 2 green ones, and 1 blue one.
>
> **_Option 1:_** You pull a cube out of the bag, record the color, and then return it to the bag. Then you pick another cube. What is the probability that the cubes are the same color?
>
> **_Option 2:_** You pull a cube out of the bag, record the color, and do not return it to the bag. Then you pick another cube. What is the probability that the cubes are the same color?

It might be easier for students to calculate the probability in **_Option 1_** than in **_Option 2_**. In **_Option 1_**, students can draw a traditional tree diagram to observe that the total number of branches is 36 and that there are 14 branches that involve cubes of the same color.

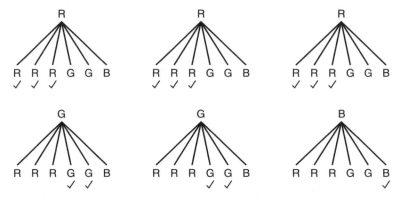

Students may be less sure of how to proceed for **_Option 2_**. They need to realize that what happens on the second choice depends in part on what happened with the first choice. Their tree diagram is not the normal symmetric type; in this case, the letters for the second choice are different groups each time.

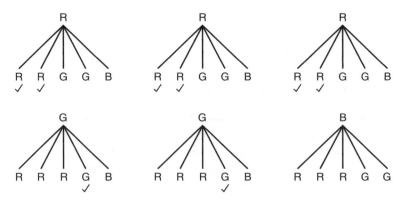

For example, if a red is chosen first, the possible colors for the second pick are 2 reds, 2 greens, and 1 blue, but if a blue is chosen first, the possible colors for the second pick are 3 reds, 2 greens, and 0 blues.

Whichever option was selected, a teacher could ask:

- *Are you more likely to draw two cubes of the same color or two cubes of different colors? Why?*
- *Is the probability of drawing two cubes of the same color more or less than $\frac{1}{2}$? How do you know?*
- *If you drew a tree diagram to represent your situation, what would it look like?*
- *What is the probability of choosing two cubes of the same color? How do you know?*

> **Option 1:** You roll two dice. Is it more likely that the sum is 8 or that the difference is 2?
>
> **Option 2:** You roll two dice. You want an event that is only a bit less likely than rolling a difference of 2. What could it be?

In either case, students must consider the probabilities associated with rolling two dice and might begin by creating a table or chart to list all of the possible outcomes. Some students will be more comfortable with **Option 1**; because the two events are listed, little decision making is required. Other students, who want to be more creative or are more comfortable with ambiguity, might prefer **Option 2**, where they must interpret what "a bit less" means and get to choose the second event.

Whichever option was selected, a teacher could ask:

- *How likely is it that the two rolls of a die differ by 2?*
- *What would the probability of an event need to be for it to be greater than the probability that the rolls differ by 2?*
- *Is the probability of rolling a sum of 8 less than the probability of rolling a difference of 2?*
- *How did you solve the problem?*

✳ **BIG IDEA.** Sometimes a probability can be estimated by using an appropriate model and conducting an experiment.

> **Option 1:** To estimate the probability that a new baby will be a boy, Tara rolls a die. She says that if she rolls a 1, 2, or 3, she will predict a boy, and if she rolls a 4, 5, or 6, she will predict a girl. Using Tara's method, estimate the probability that, for each of the next 10 days, the first two babies born in Chicago will be boys?
>
> **Option 2:** Describe and use a method to estimate the probability that a ball player who is batting 0.250 (getting 1 hit every 4 times at bat) will get 3 or more hits the next five times he comes up to bat?

Both of the suggested tasks require students to use a **simulation**. They involve using a model that is a reasonable representation of a real-life situation to estimate a probability that would be very difficult to determine otherwise.

In *Option 1*, the model is described for the student. The student then needs to figure out exactly how to set up the experiment. In *Option 2*, the student must determine both the model and the method for conducting the simulation; in that sense, *Option 2* requires a bit more of the student.

In either case, a student could be asked:

- *Why does your model for the real-life situation you are describing make sense?*
- *How did you set up your experiment?*
- *Why did it make sense to set it up that way?*
- *How many trials did you do?*
- *Do you think there were enough trials?*
- *What is your estimated probability?*
- *Does that estimate make sense to you? Why or why not?*

SUMMING UP

MY OWN QUESTIONS AND TASKS

Lesson Goal: Grade Level: _____

Standard(s) Addressed:

Underlying Big Idea(s):

_____

Open Question(s):

_____

Parallel Tasks:
Option 1:

Option 2:

_____

Principles to Keep in Mind:
- All open questions must allow for correct responses at a variety of levels.
- Parallel tasks need to be created with variations that allow struggling students to be successful and proficient students to be challenged.
- Questions and tasks should be constructed in such a way that will allow all students to participate together in follow-up discussions.

The seven big ideas that underpin work in Data Analysis and Probability were explored in this chapter through more than 40 examples of open questions and parallel tasks, as well as variations of them. The instructional examples provided were designed to support differentiated instruction for students at different developmental levels, targeting three separate grade bands: pre-K–grade 2, grades 3–5, and grades 6–8.

The examples presented in this chapter only scratch the surface of possible questions and tasks that can be used to differentiate instruction in Data Analysis and Probability. Surely many new ideas have already come to mind. For example, students could sort types of items or numbers other than those proposed, they could create graphs of other data, or they could compare and contrast different types of graphs. A form such as the one shown here can serve as a convenient template for creating your own open questions and parallel tasks. The Appendix includes a full-size blank form and tips for using it to design customized teaching materials.

Conclusions

VIRTUALLY EVERY TEACHER of mathematics in the primary, elementary, or middle-school grades is faced with students who vary in mathematical readiness for what is being taught. The differences among students may be cognitive or may relate to learning style or preferences. Regardless of the source or nature of these differences, every teacher wants to help each student succeed in learning. One of the best approaches to fostering this success within today's classroom environment is through differentiating instruction. This book was written to help teachers accomplish this differentiation.

THE NEED FOR MANAGEABLE STRATEGIES

Realistically, teachers do not have the preparation time to develop or the instructional time to deliver alternative programs that specifically target the individual learning levels of the many students with whom they work. What is needed is a way for teachers to engage all students at an appropriate level with a single question or set of related tasks. This book has modeled two core strategies that could be all a teacher requires to meet the mathematical learning needs of most students. These strategies are open questions and parallel tasks.

Open Questions

A question is open when it is framed in such a way that a variety of responses or approaches are possible. To be useful in instruction, it is important that the question focus on an important mathematical concept or skill, that the teacher has considered the readiness of most of his or her students to deal with the question at some level, and that the teacher is comfortable accepting a broad range of appropriate responses from class members.

Parallel Tasks

Two or more tasks are parallel when they are designed to meet the needs of students at different development levels, yet address the same big idea and are close enough in context that they can be discussed simultaneously. In implementing a lesson with parallel tasks, it is important that the teacher be prepared with a series of at least two or three follow-up questions about student work that would be pertinent no matter which task a student had completed. Sometimes the teacher should decide which of the tasks each student is to complete, but at other times the students should be allowed to choose.

DEVELOPING OPEN QUESTIONS AND PARALLEL TASKS

The same general approaches can be applied to developing open questions and parallel tasks for mathematical instruction.

A Three-Step Process

Step 1: Review Examples. This book provides more than 250 open questions and parallel tasks that can serve as models for new questions and tasks. A review of the examples for a particular content strand at a particular grade band will give a good sense of appropriate instructional material. By using the examples, and the variations proposed, a teacher will have a firm foundation upon which to build.

Step 2: Create the Question or Tasks. The form provided in the Appendix can serve as a convenient template for creating new open questions and parallel tasks. Among other things, the form offers a reminder that coherent instruction is served through a focus on big ideas as a means of addressing content standards. Bearing in mind how students differ with respect to specific behaviors related to the targeted big idea, the teacher can create original questions and tasks or adapt them from an existing resource. The objective in all cases should be to produce materials that appropriately challenge a broad range of students, from those who are struggling to those who are "average" to those who are particularly capable.

Step 3: Plan for Follow-Up. The teacher must ensure that the conversation resulting from class work with an open question or set of parallel tasks is inclusive. Careful structuring of questions and coordination of tasks is crucial. In follow-up discussion, the teacher should try to build connections among the various levels of responses that students provide. Making these connections clear will help the struggling student see how the answer that he or she provided can be an important step toward a more sophisticated answer.

Creating an Open Question: An Example

Consider, for example, a situation in which the intent is to teach division of three-digit numbers by one-digit numbers in a classroom where many students are still struggling with their multiplication facts. Those struggling students might not be ready for the types of questions that were initially planned. To adapt the lesson for a differentiated approach, the teacher might choose a straightforward question from a text, such as:

Suppose 4 students were delivering 176 newspapers and decided to share the task fairly. How many papers would each deliver?

Recognizing that some students were not ready for this question, the teacher might ask, instead:

Choose a number of newspapers to be delivered and a number of students to deliver them. The job should be shared fairly. How many papers should each deliver?

The change is simple but important. Now every student can work at a level that benefits him or her. In the discussion of the problem as students share their work, every student in the class (no matter what numbers that person chose) gets to hear how other students handled the situation with larger or smaller numbers. A win-win!

Creating a Set of Parallel Tasks: An Example

Consider, again, the case of the division example posed earlier. It might be that the original problem involving the 176 newspapers is posed, along with an alternate problem in which two students share the work of delivering 24 newspapers. Students might be directed to a specific problem or offered a choice of which problem to complete. However, it is critical that the discussion of the problems be common to both groups. For example, the teacher could ask:

* *What operation did you do to decide how many papers each student would deliver?*
* *Why would you use that operation?*
* *Is there another way you could have determined the answer?*
* *How did you know that each student had to deliver more than 10 papers?*
* *How do you know that each student had to deliver fewer than 100 papers?*
* *How did you figure out how many papers each student had to deliver?*

There are two important points to note: (1) The context being the same is important to making a common discussion possible. (2) It is also important to set up the situation so that there could be common questions beyond simply *What did you do?*

Fail-Safe Strategies

Open questions can be created relatively easily by strategies such as giving an answer and asking for a question for which it is the answer, allowing students to choose numbers within a question, or asking how two items are alike and different.

Parallel tasks can be generated from a single original task by strategies such as changing the complexity of the numbers, shapes, graphs, patterns, equations, or measurements being employed or the complexity of the situations being addressed.

THE BENEFITS OF THESE STRATEGIES

Teachers may at first find it challenging to differentiate their teaching by incorporating open questions and parallel tasks. Before long, however, thinking about ways to open up instruction will become a habit when content that might be too difficult for some students must be taught. The techniques illustrated in this book will become valuable tools for adjusting instruction to accommodate students at all developmental levels.

In the beginning, it might make sense for a teacher to commit to using open questions at least once a week, ultimately aiming to use such questions every one or two class periods. And a teacher might commit to using parallel tasks at least once a week, working toward employing them even more frequently. Eventually, students might even be encouraged to create their own parallel tasks by changing the complexity of situations they are asked to address.

The use of these straightforward but powerful differentiation strategies will lead to a classroom with much higher engagement levels, much less frustration, and much more math learning, thinking, and talking—promoting a much more positive attitude toward math for most students.

Appendix:
Worksheet for Open Questions
and Parallel Tasks

STRATEGIES for developing open questions and parallel tasks are discussed in Chapter 1 and in the Conclusions. The form that appears on the next page can serve as a convenient template for creation of customized materials to support differentiation of instruction in math. The form is also available for download on the Teachers College Press website: www.tcpress.com.

The following fundamental principles should be kept in mind when developing new questions and tasks:

- All open questions must allow for correct responses at a variety of levels.
- Parallel tasks need to be created with variations that allow struggling students to be successful and proficient students to be challenged.
- Questions and tasks should be constructed in such a way that all students can participate together in follow-up discussions.

MY OWN QUESTIONS AND TASKS

Lesson Goal: **Grade Level:** _____

Standard(s) Addressed:

Underlying Big Idea(s):

Open Question(s):

Parallel Tasks:

Option 1:

Option 2:

Principles to Keep in Mind:

- All open questions must allow for correct responses at a variety of levels.
- Parallel tasks need to be created with variations that allow struggling students to be successful and proficient students to be challenged.
- Questions and tasks should be constructed in such a way that will allow all students to participate together in follow-up discussions.

Glossary

For each Glossary entry, the chapter and page of first occurrence of the term are given in brackets at the end of the definition.

100 chart. A chart with the numbers from 1 to 100 displayed in order in 10 rows of 10 [Chap. 2, p. 23]

acute triangle. A triangle with three angles less than 90° [Chap. 3, p. 73]

algebraic expression. A combination of variables, operation signs, and numbers (e.g., $2n + 4$) [Chap. 5, p. 122]

algorithm. A standard procedure for performing a task (e.g., a standard procedure for adding numbers) [Chap. 2, p. 16]

array. A rectangular arrangement of objects in rows of equal length and columns of equal height [Chap. 2, p. 44]

attribute. A feature of an object (e.g., one of the attributes of a triangle is the relationship between its side lengths) [Chap. 3, p. 61]

attribute blocks. A set of blocks that vary by shape, color, size, and sometimes thickness [Chap. 6, p. 152]

bar graph. A way to show and compare data by using horizontal or vertical bars [Chap. 6, p. 149]

base ten blocks. Blocks of different sizes representing place value columns; a larger block is 10 times the size of the next smaller block [Chap. 2, p. 22]

baseline. A vertical or horizontal starting line used so that length measurements can be compared [Chap. 4, p. 95]

benchmark. Familiar measurements used for comparing other measurements (e.g., 1 inch, 1 foot, the width of a hand, the length of a foot) [Chap. 4, p. 94]

benchmark numbers. Familiar numbers used as referents for other numbers (e.g., 25 is a benchmark for numbers such as 22, 27, or 28) [Chap. 1, p. 4]

big idea. A statement that is fundamental to mathematical understanding; usually links many specific math outcomes [Chap. 1, p. 4]

box and whisker plot. A graph designed to show how data are spread out; above a number line the minimum and maximum points are plotted and connected, and the middle 50% of the data are boxed in [Chap. 6, p. 150]

cardinal number. A number that describes how many (e.g., the cardinal number 7 describes the number of days in a week) [Chap. 2, p. 19]

circumference. The distance around a circle [Chap. 4, p. 94]

coefficient. A numerical value that is the multiplier of a term involving a variable in an algebraic expression [Chap. 5, p. 146]

composite numbers. Whole numbers with more than two factors (e.g., 6 is composite because it has factors of 1, 2, 3, and 6) [Chap. 2, p. 55]

compound events. Two or more events that must both be considered in establishing a probability [Chap. 6, p. 150]

concrete graph. A way to show and compare data by using actual objects lined up horizontally or vertically [Chap. 6, p. 151]

congruent. Being of the same size and shape [Chap. 3, p. 60]

conjecture. A hypothesis or guess that something is true [Chap. 2, p. 35]

constant. A value that does not change when the value for the variable in an algebraic expression changes [Chap. 5, p. 146]

coordinate grid. A grid made up of vertical and horizontal intersecting lines; positions on the grid can be located by using two numbers—one describing the horizontal distance from a designated point and one describing the vertical distance from that point [Chap. 3, p. 60]

core. One copy of the part of a repeating pattern that repeats [Chap. 5, p. 124]

cross-section. The two-dimensional shape resulting from cutting through a three-dimensional shape with a flat plane [Chap. 3, p. 75]

Cuisenaire rods. A set of 10 colored rods of lengths 1 cm, 2 cm, 3 cm, . . . , 10 cm [Chap. 3, p. 66]

decreasing pattern. A pattern in which the values keep decreasing [Chap. 5, p. 122]

denominator. The number on the bottom of a fraction that tells the number of equal parts into which the whole is divided [Chap. 2, p. 28]

diagonal. A line segment connecting points in a polygon that are not already connected [Chap. 3, p. 74]

differentiating instruction. Tailoring instruction to meet the needs of different types or levels of students in a classroom [Chap. 1, p. 1]

dividend. The number being shared or grouped in a division situation (e.g., if 9 is divided by 3, 9 is the dividend) [Chap. 2, p. 36]

divisor. The group size or number in a group in a division situation (e.g., if 8 is divided by 2, 2 is the divisor) [Chap. 2, p. 36]

double bar graph. A bar graph in which two sets of data are displayed simultaneously by using the same categories and scale [Chap. 6, p. 150]

edge. A line segment formed where two faces of a three-dimensional shape meet [Chap. 3, p. 60]

equidistant. Equally far apart [Chap. 3, p. 91]

equilateral triangle. A triangle with three equal sides [Chap. 3, p. 73]

equivalent equation. An equation with exactly the same solutions as a given one (e.g., $2x = 4$ is equivalent to $2x + 1 = 4 + 1$) [Chap. 5, p. 125]

equivalent fractions. Two different fractions representing the same part of a whole (e.g., $\frac{1}{4}$ and $\frac{2}{8}$) [Chap. 2, p. 16]

experimental probability. The probability that results from an experiment [Chap. 6, p. 150]

exponent. The symbol in a power expression that describes how many times a factor is multiplied by itself (e.g., 4 is the exponent in 3^4, indicating that 3 is multiplied by itself 4 times: $3 \times 3 \times 3 \times 3$) [Chap. 2, p. 16]

face. A flat side of a three-dimensional shape (e.g., cubes have six faces that are all squares) [Chap. 3, p. 60]

factor. One of the whole numbers that is multiplied to produce another number (e.g., 4 is a factor of 8, because $2 \times 4 = 8$) [Chap. 2, p. 35]

flip line. The line on which a mirror is placed to create a reflection [Chap. 3, p. 72]

geoboard. A plastic grid, usually square, with pegs at intersection points; rubber bands can be placed on the pegs to create shapes [Chap. 3, p. 67]

glyph. A visual/pictorial representation of information in which different parts of the picture represent different pieces of data [Chap. 6, p. 168]

hexagon. A shape with six straight sides [Chap. 3, p. 82]

histogram. A continuous bar graph; bars are joined at the sides because the data they represent are continuous [Chap. 6, p. 150]

hypotenuse. The longest side of a right triangle [Chap. 3, p. 72]

increasing pattern. A pattern in which the values keep increasing [Chap. 5, p. 122]

interval. The distance between two endpoints on a graph scale [Chap. 6, p. 171]

isometric drawing. A two-dimensional representation of a three-dimensional shape in which equal lengths on the three-dimensional shape are drawn as equal lengths on the two-dimensional representation [Chap. 3, p. 89]

isosceles trapezoid. A trapezoid with two nonparallel sides of equal length [Chap. 3, p. 75]

isosceles triangle. A triangle with two sides of equal length [Chap. 3, p. 67]

iterative rule. A rule that tells how to start a pattern and how to get from one term to the next [Chap. 5, p. 143]

line graph. A graph designed to reveal a trend; data points are usually connected by line segments or curves [Chap. 6, p. 150]

line plot. A way to show and compare data that can be sorted into numerical categories; formed by making marks in a column above positions on the number line for each category [Chap. 6, p. 150]

line symmetry. A property of shapes in which one side is a mirror reflection of the other side [Chap. 3, p. 60]

linear pattern. A pattern in which the numbers increase by the same amount each time [Chap. 5, p. 128]

locus. A set of points that meets a specified condition (e.g., the locus of points equally far from a single point is a circle) [Chap. 3, p. 91]

mean. A value for a set of numbers determined by adding them and dividing by the number of numbers in the set; average [Chap. 6, p. 150]

median. The middle number in an ordered set of numbers [Chap. 6, p. 150]

multiple. A number that is the product of two other whole numbers (e.g., 6 is a multiple of 3) [Chap. 2, p. 35]

negative integers. The opposites of the counting numbers 1, 2, 3, ... on the other side of 0 on the number line (i.e., $-1, -2, -3, \ldots$) [Chap. 2, p. 16]

net. A two-dimensional shape that can be folded to create a three-dimensional object [Chap. 4, p. 103]

nonstandard unit. A measurement used as the basis for describing other measurements, but one that may vary from person to person (e.g., the length of a pencil or the width of a hand might be used as units for describing length) [Chap. 4, p. 97]

number line. A diagram that shows ordered numbers as points on a line [Chap. 2, p. 20]

numerator. The number on the top of a fraction that tells the number of parts of a whole under consideration [Chap. 2, p. 28]

obtuse triangle. A triangle with one angle greater than 90° [Chap. 3, p. 73]

open questions. Broad-based questions with many appropriate responses or approaches; the questions serve as a vehicle for differentiating instruction if the range of possibilities allows students at different developmental levels to succeed even while responding differently [Chap. 1, p. 6]

operation rule. A rule describing what operations to perform on a pair of numbers to get a resulting output [Chap. 5, page 146]

opposites. Two numbers on a number line that are the same distance from 0 (e.g., 8 and −8) [Chap. 2, p. 25]

ordinal number. A number describing a position in an ordered series (e.g., seventh is an ordinal number) [Chap. 2, p. 19]

pan balance. A tool used to compare weights; when an item is placed on each side of the pan balance, if the pans are at the same level, then the weights are the same [Chap. 4, p. 110]

parallel tasks. Sets of two or more related tasks or activities that explore the same big idea but are designed to suit the needs of students at different developmental levels; the tasks are similar enough in context that they can be considered together in class discussion [Chap. 1, p. 6]

parallelogram. A quadrilateral with opposite sides that are parallel [Chap. 3, p. 65]

pattern block triangles. Small equilateral triangles found in a set of shapes that also includes hexagons, trapezoids, squares, and parallelograms of particular sizes [Chap. 3, p. 70]

pattern blocks. A set of shapes that includes hexagons, trapezoids, squares, parallelograms, and triangles of particular sizes [Chap. 4, p. 100]

pattern rule. An iterative rule or relationship rule describing how the terms of a pattern are determined (e.g., the pattern 4, 7, 10, 13, . . . can be described by the rule "starts at 4 and increases by 3 each time" or by the rule "triple the position number and add 1") [Chap. 2, p. 55]

pentagon. A polygon with five sides [Chap. 3, p. 84]

perimeter. The distance around a shape [Chap. 3, p. 66]

perpendicular distance. The distance from a point to a line that is measured perpendicular to the given line [Chap. 3, p. 84]

pictograph. A graph that uses symbols to represent quantities [Chap. 6, p. 158]

picture graph. A graph that uses pictures to represent quantities [Chap. 6, p. 149]

place value system. A system used to represent numbers in which the position of a symbol affects its value (e.g., the 2 in 23 is worth a different amount than the 2 in 217) [Chap. 2, p. 16]

polydrons. Plastic shapes that can be connected to form three-dimensional shapes [Chap. 3, p. 82]

polygon. A closed shape made of only straight sides [Chap. 3, p. 68]

polyhedron. A three-dimensional shape with only polygon faces; the plural is *polyhedra* [Chap. 3, p. 80]

position number. A number that tells the position of a term in a pattern [Chap. 5, p. 132]

positional vocabulary. Words used to describe relative position (e.g., *above, below, next to*) [Chap. 3, p. 59]

power of 10. The number 10 or the result of multiplying 10 by itself repeatedly (e.g., 10, 100, 1,000, 10,000) [Chap. 2, p. 36]

prime numbers. Counting numbers with no whole-number factors other than themselves and 1 (e.g., 17 is a prime number because its only factors are 1 and 17) [Chap. 2, p. 55]

prism. A three-dimensional shape with two identical polygon bases that are parallel and opposite each other [Chap. 3, p. 63]

product. The result of multiplication (e.g., in the expression $5 \times 3 = 15$, 15 is the product) [Chap. 2, p. 16]

properties. An attribute that applies to all items in a group (e.g., a property of squares is that they have four identical corners) [Chap. 3, p. 60]

proportion. A description of how two numbers are related based on multiplication (e.g., the proportion 25:100 can also be described as the fraction $\frac{1}{4}$, the decimal 0.25, or the ratio 1:4) [Chap. 2, p. 16]

proportional thinking. A focus on comparing numbers based on multiplication (e.g., thinking of 12 as four 3s rather than as, e.g., $10 + 2$) [Chap. 4, p. 112]

pyramid. A shape with a polygon base and triangular faces that meet at a point [Chap. 3, p. 75]

Pythagorean theorem. A statement that the square of the length of the longest side of a right triangle is the sum of the squares of the lengths of the two other sides [Chap. 3, p. 60]

quadrant. One fourth of a coordinate grid; the first quadrant is to the right and above the center point; the second quadrant is to the left and above the center point; the third quadrant is to the left and below the center point; and the fourth quadrant is to the right and below the center point [Chap. 3, p. 79]

quadrilateral. A four-sided polygon [Chap. 3, p. 62]

quotient. The result of division (e.g., in the expression $8 \div 4 = 2$, 2 is the quotient) [Chap. 2, p. 16]

rational numbers. Numbers that can be expressed as the quotient of two integers (e.g., 0.25, -4.5, or $\frac{3}{11}$) [Chap. 2, p. 34]

rectangular prism. A prism with rectangle-shaped bases [Chap. 3, p. 62]

reflection. The mirror image of (or result of flipping) a shape [Chap. 3, p. 78]

reflection line. A flip line [Chap. 3, p. 78]

regrouping. Trading tens, hundreds, and so on for the next lower level of place value, or vice versa, usually while performing a calculation (e.g., rewriting 56 as 40 + 16 to solve 56 − 9) [Chap. 2, p. 30]

regular hexagon. A shape with six equal straight sides [Chap. 3, p. 82]

relationship rule. A rule that tells how to determine the value of a term in a pattern if its location in the pattern is known [Chap. 5, p. 144]

repeating decimals. Decimals with a set of repeating digits that go on forever (e.g., 0.121212 . . . is a repeating decimal, as is 0.45979797979797 . . .); repeating decimals cannot be expressed as tenths, hundredths, thousandths, and so on [Chap. 2, p. 33]

repeating pattern. A pattern in which a certain number of terms are repeated over and over [Chap. 5, p. 122]

rhombus. A parallelogram with four sides of equal length [Chap. 4, p. 101]

right angle. A 90° angle [Chap. 3, p. 73]

right trapezoid. A trapezoid with a 90° angle [Chap. 3, p. 76]

right triangle. A triangle with a 90° angle [Chap. 3, p. 73]

rotational symmetry. An attribute of shapes that can be turned less than a full turn to fit into their own outline [Chap. 3, p. 60]

scale. The amount that one unit in a graph represents (e.g., if each square in a bar graph represents 5, the scale is 5) [Chap. 6, p. 158]

scalene triangle. A triangle with all sides of unequal lengths [Chap. 3, p. 73]

scatterplot. A set of points plotted on a coordinate grid designed to make relationships between two variables evident [Chap. 6, p. 150]

scientific notation. A way of describing numbers as the product of a number between 1 and 10 and a power of 10 (e.g., 34.5 can be written as 3.45×10^1 in scientific notation) [Chap. 2, p. 16]

similarity. A relationship between two shapes such that one is the same shape but of a different size than the other; in effect, an enlargement or reduction [Chap. 3, p. 60]

simulation. A model that can be used to mimic a real-life situation [Chap. 6, p. 180]

skeleton. A representation of a shape that uses only its edges and vertices [Chap. 3, p. 70]

slide. A translation [Chap. 3, p. 86]

snap cubes. Plastic cubes that can be linked together [Chap. 5, p. 130]

sorting rule. A description of how it is determined which items of a group belong together [Chap. 6, p. 149]

square number. A number that is the product of a whole number multiplied by itself [Chap. 4, p. 103]

square root. A number that can be multiplied by itself to result in a given number (e.g., the square root of 16 is 4, because $4 \times 4 = 16$) [Chap. 2, p. 33]

standard measurement. An unambiguous measurement based on units that hold the same meaning for anyone (e.g., inch, mile, square meter) [Chap. 4, p. 94]

stem and leaf plot. A way to organize data in groups based on place value (e.g., 40s, 50s, 60s) [Chap. 6, p. 150]

substitution. Replacement of a variable with a given number [Chap. 5, p. 147]

summary statistic. A single number that in some way represents an entire set of data (e.g., mean) [Chap. 6, p. 150]

surface area. The total of the area of the surfaces of a three-dimensional object [Chap. 4, p. 94]

table of values. A way to organize related information in a chart with two columns [Chap. 5, p. 132]

tally chart. A set of small lines organized in groups of five to show a count [Chap. 6, p. 154]

tangram. A puzzle in the form of a square made up of seven specific pieces [Chap. 3, p. 71]

term (value). The number or object that is found in a particular location in a pattern [Chap. 5, p. 124]

terms (of a pattern). The values that form a pattern (e.g., the terms of the pattern 10, 20, 30, . . . are the numbers 10, 20, 30, 40, . . .) [Chap. 2, p. 55]

theoretical probability. The probability that would be expected when analyzing all of the possible outcomes of a situation [Chap. 6, p. 150]

thousandths grid. A grid made up of 10 rows and 10 columns wherein each of the 100 squares in the grid is divided into 10 small rectangles; altogether there are 1,000 small rectangles in the grid [Chap. 2, p. 52]

transformation. A motion that moves a shape from one position to another; translations, reflections, and rotations are common motions studied [Chap. 3, p. 60]

translation. A slide with a vertical and/or horizontal component [Chap. 3, p. 79]

transparent mirror. A mirror that can be seen through; with this type of mirror, a student can draw the image of an object that is viewed in it [Chap. 3, p. 72]

trapezoid. A quadrilateral with one pair of parallel sides [Chap. 3, p. 76]

tree diagram. A way to record and count all combinations of events by using lines to form branches [Chap. 6, p. 162]

triangle-based prism. A prism in which the two opposite bases are triangles [Chap. 3, p. 84]

triangle-based pyramid. A pyramid with a triangle base [Chap. 3, p. 75]

unit. A measurement used as the basis for describing other measurements (e.g., inch is a unit of length) [Chap. 4, p. 93]

unit fractions. Fractions with a numerator of 1 [Chap. 2, p. 28]

variable. A quantity that varies or changes or is unknown; often represented by an open box or a letter [Chap. 5, p. 122]

vertex. A corner of a shape; the plural is *vertices* [Chap. 3, p. 60]

x-coordinate. The first in a pair of numbers that describes a point's horizontal distance from a reference point on a coordinate grid [Chap. 5, p. 132]

y-coordinate. The second in a pair of numbers that describes a point's vertical distance from a reference point on a coordinate grid [Chap. 5, p. 132]

zone of proximal development. The range of potential learning that is beyond existing knowledge but that is accessible to a student with adult or colleague support [Chap. 1, p. 3]

Bibliography

Anderson, M., & Little, D. M. (2004). On the write path: Improving communication in an elementary mathematics classroom. *Teaching Children Mathematics, 10,* 468–472.

Bamberger, H., & Hughes, P. (1999). *Super graphs, Venns & glyphs.* New York: Scholastic.

Barrett, J. E., Jones, G., Thornton, C., & Dickson, S. (2003). Understanding children's developing strategies about concepts for length. In D. H. Clement & G. Bright (Eds.), *Learning and teaching measurement* (pp. 17–30). Reston, VA: National Council of Teachers of Mathematics.

Borgoli, G. M. (2008). Equity for English language learners in mathematics classrooms. *Teaching Children Mathematics, 15,* 185–191.

Bremigan, E. G. (2003). Developing a meaningful understanding of the mean. *Mathematics Teaching in the Middle School, 9,* 22–27.

Bright, G. W., Brewer, W., McClain, K., & Mooney, E. S. (2003). *Navigating through data analysis in grades 6–8.* Reston, VA: National Council of Teachers of Mathematics.

Cavanagh, M., Dacey, L., Findell, C. R., Greenes, C. E., Sheffield, L. J., & Small, M. (2004). *Navigating through number and operations in prekindergarten–grade 2.* Reston, VA: National Council of Teachers of Mathematics.

Charles, R. (2004). Big ideas and understandings as the foundation for elementary and middle school mathematics. *Journal of Mathematics Education Leadership, 7,* 9–24.

Clements, D. H. (1999). Geometric and spatial thinking in young children. In J. Copley (Ed.), *Mathematics in the early years* (pp. 66–79). Washington, DC: National Association for the Education of Young Children and National Council of Teachers of Mathematics.

Cuevas, G. J., & Yeatts, K. (2001). *Navigating through algebra, grades 3–5.* Reston, VA: National Council of Teachers of Mathematics.

Dacey, L., Cavanagh, M., Findell, C. R., Greenes, C. E., Sheffield, L. J., & Small, M. (2003). *Navigating through measurement in prekindergarten–grade 2.* Reston, VA: National Council of Teachers of Mathematics.

Dacey, L., & Lynch, J. B. (2007). *Math for all: Differentiating instruction, grades 3–5.* Sausalito, CA: Math Solutions Publications.

Dacey, L., & Salemi, R. E. (2007). *Math for all: Differentiating instruction, grades K–2.* Sausalito, CA: Math Solutions Publications.

Drake, J. M., & Barlow, A. T. (2007). Assessing students' levels of understanding multiplication through problem writing. *Teaching Children Mathematics, 14,* 272–277.

Findell, C. R., Small, M., Cavanagh, M., Dacey, L., Greenes, C. E., & Sheffield, L. J. (2001). *Navigating through geometry in prekindergarten–grade 2.* Reston, VA: National Council of Teachers of Mathematics.

Forman, E. A. (2003). A sociocultural approach to mathematics reform: Speaking, inscribing, and doing mathematics within communities of practice. In J. Kilpatrick, W. G. Martin, & D. Schifter (Eds.), *A research companion to* Principles and Standards for School Mathematics (pp. 333–352). Reston, VA: National Council of Teachers of Mathematics.

Friel, S., Rachlin, S., & Doyle, D. (2001). *Navigating through algebra, grades 6–8.* Reston, VA: National Council of Teachers of Mathematics.

Greenes, C. E., Cavanagh, M., Dacey, L., Findell, C. R., & Small, M. (2001). *Navigating through algebra in prekindergarten–grade 2.* Reston, VA: National Council of Teachers of Mathematics.

Gregory, G. H., & Chapman, C. (2006). *Differentiated instructional strategies: One size doesn't fit all* (2nd ed.). Thousand Oaks, CA: Corwin Press.

Imm, K. L., Stylianou, D. A., & Chae, N. (2008). Student representations at the center: Promoting classroom equity. *Mathematics Teaching in the Middle School, 13,* 458–463.

Jones, G. A., & Thornton, C. A. (1993). Children's understanding of place value: A framework for curriculum development and assessment. *Young Children, 48,* 12–18.

Karp, K., & Howell, P. (2004). Building responsibility for learning in students with special needs. *Teaching Children Mathematics, 11,* 118–126.

Kersaint, G. (2007). The learning environment: Its influence on what is learned. In W. G. Martin, M. E. Strutchens, & P. C. Elliott (Eds.), *The learning of mathematics* (pp. 83–96). Reston, VA: National Council of Teachers of Mathematics.

Kieran, T., Davis, B., & Mason, R. (1996). Fraction flags: Learning from children to help children learn. *Mathematics Teaching in the Middle School, 2,* 14–19.

Lovin, A., Kyger, M., & Allsopp, D. (2004). Differentiation for special needs learners. *Teaching Children Mathematics, 11,* 158–167.

Martin, H. (2006). *Differentiated instruction for mathematics.* Portland, ME: Walch.

Mohr, D. J., Walcott, C. Y., & Kastberg, S. E. (2008). Using your inner voice to guide instruction. *Teaching Children Mathematics, 15,* 112–119.

Moyer, P. S. (2001). Using representations to explore perimeter and area. *Teaching Children Mathematics, 8,* 52–59.

Murray, M., & Jorgensen, J. (2007). *The differentiated math classroom: A guide for teachers, K–8.* Portsmouth, NH: Heinemann.

National Council of Teachers of Mathematics. (2000). *Principles and standards for school mathematics.* Reston, VA: National Council of Teachers of Mathematics.

National Council of Teachers of Mathematics. (2006). *Curriculum focal points.* Reston, VA: National Council of Teachers of Mathematics.

O'Connell, S. (2007a). *Introduction to communication, grades preK–2.* Portsmouth, NH: Heinemann.

O'Connell, S. (2007b). *Introduction to communication, grades 3–5.* Portsmouth, NH: Heinemann.

Pugalee, D. K., Frykholm, J., Johnson, A., Slovin, H., Malloy, C., & Preston, R. (2002). *Navigating through geometry in grades 6–8.* Reston, VA: National Council of Teachers of Mathematics.

Sakshaug, L. (2000). Which graph is which? *Teaching Children Mathematics, 6,* 454–455.

Schifter, D., Bastable, V., & Russell, S. I. (1997). Attention to mathematical thinking: Teaching to the big ideas. In S. Friel & G. Bright (Eds.), *Reflecting on our work: NSF teacher enhancement in mathematics K–6* (pp. 255–261). Washington, DC: University Press of America.

Sheffield, L. J. (2003). *Extending the challenge in mathematics: Developing mathematical promise in K–8 students.* Thousand Oaks, CA: Corwin Press.

Sheffield, L. J., Cavanagh, M., Dacey, L., Findell, C. R., Greenes, C. E., & Small, M. (2002). *Navigating through data analysis and probability in prekindergarten–grade 2.* Reston, VA: National Council of Teachers of Mathematics.

Small, M. (2005a). *PRIME: Professional resources and instruction for mathematics educators: Number and operations.* Toronto: Thomson Nelson.

Small, M. (2005b). *PRIME: Professional resources and instruction for mathematics educators: Patterns and algebra.* Toronto: Thomson Nelson.

Small, M. (2006). *PRIME: Professional resources and instruction for mathematics educators: Data management and probability.* Toronto: Thomson Nelson.

Small, M. (2007). *PRIME: Professional resources and instruction for mathematics educators: Geometry.* Toronto: Thomson Nelson.

Small, M. (2008). *Making math meaningful to Canadian students: K–8.* Toronto: Nelson Education Ltd.

Tomlinson, C. A. (1999). *The differentiated classroom: Responding to the needs of all learners.* Alexandria, VA: Association for Supervision and Curriculum Development.

Tomlinson, C. A. (2001). *How to differentiate instruction in a mixed ability classroom* (2nd ed.). Alexandria, VA: Association for Supervision and Curriculum Development.

Tomlinson, C. A., & McTighe, J. (2006). *Integrating differentiated instruction and understanding by design.* Alexandria, VA: Association for Supervision and Curriculum Development.

Tompert, A. (1990) *Grandfather Tang's story.* New York: Crown.

Vygotsky, L. S. (1978). *Mind in society: The development of higher psychological processes.* Cambridge, MA: Harvard University Press.

Westphal, L. (2007). *Differentiating instruction with menus: Math.* Austin, TX: Prufrock Press.

Williams, L. (2008). Tiering and scaffolding: Two strategies for providing access to important mathematics. *Teaching Children Mathematics, 14,* 324–330.

Index

The Index is divided into two sections: the Index of Subjects and Cited Authors and the Index of Big Ideas. The Index of Subjects and Cited Authors covers instructional concepts as well as listings for content areas and grade bands; the names of authors cited in the text are also included. The Index of Big Ideas covers mathematical content broken down by big idea, providing access to the broad concepts presented. Individual mathematical terms are not indexed. The Glossary (pages 187–195) lists the primary mathematical terms featured in the text and instructional examples. Each Glossary entry ends with a chapter and page designator identifying the location of the first occurrence of each term.

INDEX OF SUBJECTS AND CITED AUTHORS

INDEX OF BIG IDEAS

The use of standard measurement units simplifies communication about the size of objects.

> Grades 3–5, 115

Knowledge of the size of benchmarks assists in measuring.

> Prekindergarten–grade 2, 98, 109–110
> Grades 3–5, 101–102
> Grades 6–8, 104–105, 117

Measurement formulas allow us to rely on measurements that are simpler to access to calculate measurements that are more complicated to access.

> Grades 6–8, 105–107, 117–119

Algebra

A group of items form a pattern only if there is an element of repetition, or regularity, that can be described with a pattern rule.

> Prekindergarten–grade 2, 123–124, 136–137
> Grades 3–5, 127–129, 139–140
> Grades 6–8, 131–132, 143–144

Any pattern, algebraic expression, relationship, or equation can be represented in many ways.

> Prekindergarten–grade 2, 124–125
> Grades 3–5, 140–141
> Grades 6–8, 132–133

Patterns are all around us in the everyday world.

> Prekindergarten–grade 2, 126

Many number, geometry, and measurement ideas are based on patterns.

> Prekindergarten–grade 2, 138–139
> Grades 3–5, 129–130, 141–142
> Grades 6–8, 134, 144–145

Arranging information in charts and tables can make patterns easier to see.

> Prekindergarten–grade 2, 126–127
> Grades 3–5, 130–131

Variables can be used to describe relationships.

> Grades 3–5, 131, 142–143
> Grades 6–8, 134–135, 146–148

Data Analysis and Probability

Many data collection activities are based on the sorting of information into meaningful categories.

> Prekindergarten–grade 2, 151–153, 165–166
> Grades 3–5, 156–157
> Grades 6–8, 160

To collect useful data, it is necessary to decide, in advance, what source or collection method is appropriate and how to use that source or method effectively.

> Prekindergarten–grade 2, 153–155
> Grades 3–5, 157

Sometimes a large set of data can be usefully described by using a summary statistic, which is a single meaningful number that describes the entire set of data. The number might describe the values of individual pieces of data or how the data are distributed or spread out.

> Grades 3–5, 170
> Grades 6–8, 161–162, 175

Graphs are powerful data displays because they quickly reveal a great deal of information.

> Prekindergarten–grade 2, 155–156, 167–168
> Grades 3–5, 157–159, 170–172
> Grades 6–8, 175–176

An experimental probability is based on past events, and a theoretical probability is based on analyzing what could happen. An experimental probability approaches a theoretical one when enough random samples are used.

> Grades 3–5, 159–160, 173–174
> Grades 6–8, 162–164, 176–179

In probability situations, one can never be sure what will happen next. This is different from most other mathematical situations.

> Prekindergarten–grade 2, 169
> Grades 6–8, 164–165

Sometimes a probability can be estimated by using an appropriate model and conducting an experiment.

> Grades 6–8, 179–180

About the Author

MARIAN SMALL is the former Dean of Education at the University of New Brunswick. She has been a professor of mathematics education for many years and is a regular speaker on K–12 mathematics throughout Canada and the United States.

The focus of Dr. Small's work has been the development of curriculum and text materials for students and teachers of mathematics. She has been an author on seven text series at both the elementary and the secondary levels in Canada, the United States, Australia, and the country of Bhutan and a senior author on five of those series. She has served on the author team for the National Council of Teachers of Mathematics Navigation series, pre-K–2. For four years, she was the NCTM representative on the Mathcounts question writing committee for middle-school mathematics competitions throughout the United States. She is also a member of the editorial panel for the NCTM 2011 yearbook on motivation and disposition.

Dr. Small has recently completed a text for university preservice teachers and practicing teachers, *Making Math Meaningful for Canadian Students: K–8,* as well as the professional resources volumes *Big Ideas from Dr. Small: Grades 4–8* and *Big Ideas from Dr. Small: Grades K–3* published by Nelson Education Ltd.

She has led the research resulting in the creation of maps describing student mathematical development in each of the five NCTM mathematical strands for the K–8 levels and has created the associated professional development program, PRIME. She has also developed materials and provided consultation focused on working with struggling learners and on teacher questioning in mathematics classrooms.